HANDBOOK OF DIFFERENTIAL ENTROPY

HANDBOOK OF DIFFERENTIAL ENTROPY

Joseph Victor Michalowicz
Jonathan M. Nichols
Frank Bucholtz

CRC Press
Taylor & Francis Group
Boca Raton London New York

CRC Press is an imprint of the
Taylor & Francis Group, an **informa** business

A CHAPMAN & HALL BOOK

CRC Press
Taylor & Francis Group
6000 Broken Sound Parkway NW, Suite 300
Boca Raton, FL 33487-2742

First issued in paperback 2019

© 2014 by Taylor & Francis Group, LLC
CRC Press is an imprint of Taylor & Francis Group, an Informa business

No claim to original U.S. Government works

ISBN-13: 978-1-4665-8316-0 (hbk)
ISBN-13: 978-1-138-37479-9 (pbk)

Library of Congress Cataloging-in-Publication Data

Michalowicz, Joseph Victor, 1941-
  Handbook of differential entropy / Joseph Victor Michalowicz, Jonathan M. Nichols, Frank Bucholtz.
      pages cm
    "A CRC title."
    Includes bibliographical references and index.
    ISBN 978-1-4665-8316-0 (hardcover : alk. paper)
    1. Entropy. 2. Mathematical physics. I. Nichols, Jonathan Michael. II. Bucholtz, Frank. III. Title.

QC318.E57M53 2014
003'.5401515352--dc23   2013030830

Visit the Taylor & Francis Web site at
http://www.taylorandfrancis.com

and the CRC Press Web site at
http://www.crcpress.com

Contents

List of Figures

List of Tables

Preface

This book is intended as a practical introduction to the topic of differential entropy for students taking a first course in information theory. The student will undoubtedly be told that there are two basic issues in communications theory: (1) the ultimate data compression possible and (2) the ultimate transmission rate. The measure for the second quantity is channel capacity, a concept which seems natural to most students. The first, however, is measured by *entropy*, which may at first appear a rather strange concept. The student will first look at discrete random variables and will see that the entropy provides a lower limit on the number of bits needed, on the average, to represent the discrete variable. Students will learn certain properties of entropy, the first of which is that it is always greater than or equal to zero. It will also become apparent which types of distributions tend to maximize and/or minimize the entropy.

Later on, the student may be introduced to the concept of differential entropy for continuous random variables. This seems to be the natural extension to discrete entropy and the relationship between the two can be demonstrated. However, the differential entropy is a very different quantity, requiring some care in the way it is interpreted and used in practice. Unlike discrete entropy, the differential entropy can be positive or negative and must be referenced to the units of the underlying random variable. In short, while it may seem to be a simple extension of the discrete case, we have found differential entropy to be a more complex measure requiring a more careful treatment.

This book is therefore devoted to making the student more comfortable with the concept of differential entropy. We will begin with a brief review of probability theory as it provides an understanding of the core building block of entropy. We will then introduce both discrete and differential entropy and attempt to shed light on the challenges associated with interpreting and deriving the latter for various probability distributions. Finally, we will examine several application areas where the differential entropy has proven useful.

TABLE 0.1
Definition of useful functions and constants.

Glossary of Key Functions and Constants	
Gamma function	$\Gamma(x) = \int_0^\infty t^{x-1} e^{-t} dt$ for $x > 0$
Digamma function	$\Psi(x) = \frac{d}{dx} \ln[\Gamma(x)] = \frac{\Gamma'(x)}{\Gamma(x)}$
Beta function	$B(x,y) = \int_0^1 t^{x-1}(1-t)^{y-1} dt = \frac{\Gamma(x)\Gamma(y)}{\Gamma(x+y)}$
Normal Cumulative Distribution Function	$P(x) = \frac{1}{2\pi} \int\limits_{-\infty}^{x} e^{-t^2/2} dt$
Euler-Mascheroni constant	$\gamma = \lim\limits_{m \to \infty} \left[1 + \frac{1}{2} + \frac{1}{3} + \cdots + \frac{1}{m} - \ln(m) \right]$ $= 0.5772156649$ $= -\Gamma'(1) = -\Psi(1)$

1

Probability in Brief

Arguably the most important goal of any scientific discipline is to develop the mathematical tools necessary for modeling and prediction. Models that accurately describe the world around us are retained; those that do not, are discarded or modified. From a modeling perspective we typically divide the world into two types of phenomena: those that are deterministic and those that are probabilistic. Deterministic models are those that can be used to predict, with near certainty, the outcome of a given experiment. Using Newton's laws of motion, for example, we can predict the exact time it will take for a ball of lead to fall to the ground when released from a height of d meters. Newton's laws tell us that the ball, when released from rest, will accelerate at $9.81m/s^2$ so that the time to reach ground is found through basic calculus to be $\sqrt{2d/9.81}$ seconds.

On the other hand, there are certain phenomena that defy a deterministic description. We cannot predict whether or not the flipping of a fair coin, for example, will result in a "head" or a "tail."[1] Intead we use a *probabilistic* model to describe the outcome of this experiment. The goal of the probabilistic model is not to predict a specific outcome, but rather to predict the likelihood or probability of a given outcome. Understanding probabilistic modeling is essential to understanding entropy. In fact, as we will show, entropy relies on a probabilistic description of an event. We will also show that entropy is itself a predictive tool that can be used to forecast limits on rates of communication between two devices, predict the amount of compression possible in a piece of data, or even assess the degree of statistical dependence among two or more experimental observations.

In this chapter we establish the concepts of probability theory required of the remainder of this text. In doing so, we cover such topics as: probability spaces, random variables, probability distributions and statistics (e.g., moments, covariance). We also devote a small section to random processes as our applications of differential entropy, described in Chapter 6, will be applied to the output of experiments typically modeled as a sequence of random variables. This chapter will further serve as a reference for the notation used throughout this work.

---

[1]In principle if we knew the exact forces acting on the coin and the precise orientation of the coin at the time of the flip we could, in fact, use Newton's laws to accurately predict the outcome. However, this is information that must be considered unknown in any practical circumstance of interest.

As we have just pointed out, probability theory is used to model an experiment whose outcome is uncertain. We will denote the outcome of the experiment with a lowercase letter, e.g., "x." In any such experiment we can denote the set of possible outcomes ξ. In an experiment with a discrete outcome, such as the roll of a dice, this set is simply a list of those possibilities, e.g., $\xi = \{1, 2, 3, 4, 5, 6\}$. For experiments where the outcome is one number in a continuous range of values, this set can be similarly defined. Consider the reading of a voltage from a sensor, conditioned so that the response is in the ± 5 volt range. In this case $\xi = [-5, 5]$. In both experiments we can define certain *events* as subsets of ξ. In the dice example we can, for example, define the set of even numbers $\sigma_1 = \{2, 4, 6\}$, the set of numbers less than 5, $\sigma_2 = \{1, 2, 3, 4\}$, or the prime numbers, $\sigma_3 = \{1, 2, 3, 5\}$. The superset containing these events (subsets) is denoted σ and must include the null set $\emptyset = \{\}$ and the set of all possibilities ξ. In the continuous case we might have $\sigma_1 = [-5, 0]$, $\sigma_2 = [-1, 1]$, and $\sigma_3 = [1.36435]$. The events $\sigma_i \in \sigma$ (whether modeling continuous or discrete experiments) must obey

$$\text{if } \sigma_i \in \sigma, \ \sigma_i^C \in \sigma \tag{1.1a}$$

$$\sigma_1 \cup \sigma_2 \cup \cdots \in \sigma \tag{1.1b}$$

$$\sigma_1 \cap \sigma_2 \cap \cdots \in \sigma \tag{1.1c}$$

where σ_i^C is the complement of σ_i defined as $\sigma_i^C \equiv \xi - \sigma_i$. That is to say, for a collection of subsets $\sigma_i \in \sigma$, then σ is closed under the formation of complements, countable unions and countable intersections. These requirements are in place to ensure the existence of limits in developing probability theory [46]. The collection of sets obeying these properties is referred to as a σ-algebra.

We have defined a space of possible outcomes and specified notation for defining specific events among those outcomes. The final step is to define a function $Pr(\sigma_i)$ that assigns a probability to the occurence of those events. We require that this function obey

$$Pr(\sigma_i) \geq 0 \tag{1.2a}$$

$$Pr(\xi) = 1 \tag{1.2b}$$

$$Pr\left(\sum \sigma_i\right) = \sum Pr(\sigma_i) \ \text{ for } \sigma_i \in \xi \text{ and } \sigma_i \cap \sigma_j = \emptyset, \ i \neq j. \tag{1.2c}$$

The first two mathematical statements simply say that probability is a non-negative quantity and that the probability assigned to the entire space of possibilities is unity. The last statement is slightly less intuitive and states that, for non-overlapping (disjoint) sets of events, the sum of their probabilities is equal to the probability assigned to their union. In the dice example, this is easy to verify. Defining a valid $\sigma-$algebra $\sigma_1 = \{1, 2, 3, 4, 5, 6\}$, $\sigma_2 = \emptyset$, $\sigma_3 = \{1, 2\}$, $\sigma_4 = \{3, 4\}$, $\sigma_5 = \{5, 6\}$ we see that $Pr(\sigma_3 \cup \sigma_4 \cup \sigma_5) = 1$ and

$Pr(\sigma_3) + Pr(\sigma_4) + Pr(\sigma_5) = 1/3 + 1/3 + 1/3 = 1$. For any experiment with discrete outcomes it is similarly straightforward to define a $\sigma-$algebra and verify that a proper probability-assigning function exists. The situation is less clear for continuous random variables. The problem is that our $\sigma-$algebra could be comprised of an uncountable infinity of singleton values. Returning to our experiment measuring voltage on the ± 5 volt range, define the event $\sigma_1 = 0.32325$. The probability of getting *exactly* this value is zero, i.e., $Pr(\sigma_1) = 0$. In fact, the probability of getting any particular value is zero so that the sum of probabilities of the uncountable infinity of events comprising our $\sigma-$algebra is zero, i.e., $\sum_i Pr(\sigma_i) = 0$. However, these events cover the space of possible outcomes so that $Pr(\sum_i \sigma_i) = 1$. This contradiction was mathematically resolved by the extension theorem [21] which states that, so long as the $\sigma-$algebra for a continuous space of possibilities is comprised of a countable set of closed intervals that completely "tile" the space, it is a valid σ-algebra to which probabilities can be assigned.

In summary, we can probabilistically model an experiment, discrete or continuous, with the triple $(\boldsymbol{\xi}, \boldsymbol{\sigma}, Pr(\cdot))$. That is to say, we have a space of possible outcomes, subsets of those possible outcomes defined to obey (1.1), and a function with properties (1.2) that assigns a probability to each subset. This construct allows us to model a particular experimental outcome, x. A single roll of our dice, for example, will yield an integer $x \in \{1, 2, 3, 4, 5, 6\}$. We do not know a priori which of the six outcomes will be realized (as we would in a deterministic model) so we choose to treat the outcome as a *random variable* $X(s \in \boldsymbol{\xi})$. A random variable is a function, defined everywhere on the space of possible experimental outcomes. While we could choose this function to be anything we would like, a sensible choice is to simply choose $X(s) \equiv s$, that is to say, the random variable *is* a particular element of the set of possible measurements we might make. From this point forward we will simply use the notation X with the understanding that for $s \in \boldsymbol{\xi}$, $X \equiv X(s) = s$. In the dice experiment, X is an integer in the range $[1, 6]$. In our voltage example, X is a particular value on the closed interval of real numbers in $[-5, 5]$. A random variable will always be denoted with the uppercase character corresponding to the experimental outcome it is being used to model, e.g., X is used to model x.

1.1 Probability Distributions

Whether we are dealing with discrete or continuous random variables we will assume that we may (1) assign a probability to the outcome of an experiment whose outcome is uncertain and (2) define a random variable X on the space of possible outcomes of that experiment. Given this construct, we define the *cumulative distribution function* (CDF)

$$P_X(x) \equiv Pr(X \leq x) \tag{1.3}$$

as the probability assigned to the event $\sigma = \{X \leq x\}$ for real x. The notation indicates that $P_X(\cdot)$ is the distribution function for the random variable X evaluated for a specfic value of X, namely x. This function will always be bounded on the range $[0, 1]$ as $P_X(-\infty) = 0$ and $P_X(\infty) = 1$. Moreover, we have from axiom (1.2c) that

$$P_X(x + dx) = Pr(X \leq x) + Pr(x < X \leq x + dx)$$
$$= P_X(x) + Pr(x < X \leq x + dx) \geq P_X(x) \tag{1.4}$$

and therefore the CDF is a monotonically increasing function.

Now, if the random variable is defined on a continous range of possible outcomes, the CDF will also typically be continuous and we may define its derivative

$$p_X(x) \equiv \lim_{\Delta x \to 0} \frac{P_X(x + \Delta x) - P_X(x)}{\Delta x} = dP_X(x)/dx \tag{1.5}$$

as the *probability density function*, or PDF, associated with X. The PDF provides the probability of observing the random variable in a given interval dx, i.e.,

$$p_X(x)dx = Pr(x < X \leq x + dx). \tag{1.6}$$

In this text we will also consider the case where our measurement can take one of a fixed set of discrete outcomes. In this case X is used to model an observation x_m, which is one of $\mathcal{M}$ possible outcomes, i.e., $m \in [1, \mathcal{M}]$, a *discrete* random variable, in which case it makes sense to define a function

$$f_X(x_m) \equiv Pr(X = x_m) \tag{1.7}$$

which assigns a probability to the event $X = x_m$. This will be referred to as a *probability mass function* (PMF). The CDF associated with this PMF can then be defined as

$$F_X(x_M) \equiv \sum_{m=1}^{M} f_X(x_m). \tag{1.8}$$

We will henceforth use $f_X(\cdot)$, $F_X(\cdot)$ to denote a PMF and the corresponding CDF and retain $p_X(\cdot)$, $P_X(\cdot)$ to describe continuous random variables. Given our focus on differential entropy, we are mostly interested in the latter.

The CDF and PDF of a continuous random variable are therefore two equivalent ways of defining a probabilistic model and are related by integration

$$P_X(x) = \int_{-\infty}^{x} p_X(u)du. \tag{1.9}$$

As an example, consider one of the most widely used probabilistic models in science, the Normal or Gaussian distribution

$$p_X(x) = \frac{1}{\sqrt{2\pi\sigma_X^2}} e^{-\frac{1}{2\sigma_X^2}(x-\mu_X)^2} \tag{1.10}$$

defined by the parameters μ_X, σ_X. The parameter μ_X governs the location of the distribution while σ_X governs the distribution width. The PDF for this distribution is given in Figure 4.37, appearing in Chapter 4 for various μ_X, σ_X. This particular model suggests outcomes near μ_X are the most likely to occur. For $\sigma \ll 1$, those outcomes are likely confined to a narrow range about μ_X while $\sigma \gg 1$ suggests the outcomes could be highly variable, occuring over a wide range of values.

The above arguments and definitions can be extended to model experiments with multiple outcomes. For example, we may be interested in the joint event that two measurements from a sensor are each between 1 and 2 volts. If the measurements are modeled as the random variables X and Y, respectively, we are interested in assigning the probability

$$Pr(1 \le X \le 2 \text{ and } 1 \le Y \le 2). \tag{1.11}$$

This question can be answered by considering the two-dimensional analogue of Eqn. (1.3), and defining the *joint* CDF

$$P_{XY}(x,y) = Pr(X \le x \text{ and } Y \le y) \tag{1.12}$$

which assigns a probability to the event that the random variables are less than the values x and y, respectively. The question posed above can therefore be answered as

$$Pr(1 \le X \le 2 \text{ and } 1 \le Y \le 2) = P_{XY}(2,2) - P_{XY}(1,1). \tag{1.13}$$

We may continue the analogy to the univariate case and define the two-dimensional, joint PDF

$$p_{XY}(x,y) = \frac{\partial^2 P_{XY}(x,y)}{\partial x \partial y}. \tag{1.14}$$

Thus, for continuous random variables we could have alternatively answered the previous question as

$$Pr(1 \le X \le 2 \text{ and } 1 \le Y \le 2) = \int_1^2 \int_1^2 p_{XY}(x,y)dxdy. \tag{1.15}$$

A common example is the bivariate extension of (1.10)

$$p_{XY}(x,y) = \frac{1}{2\pi\sigma_X\sigma_Y\sqrt{1-\rho_{XY}^2}} e^{-\frac{1}{2(1-\rho_{XY}^2)}\left[\frac{(x-\mu_X)^2}{\sigma_X^2} + \frac{(y-\mu_Y)^2}{\sigma_Y^2} - \frac{2\rho_{XY}(x-\mu_X)(y-\mu_Y)}{\sigma_X\sigma_Y}\right]}$$

$$\tag{1.16}$$

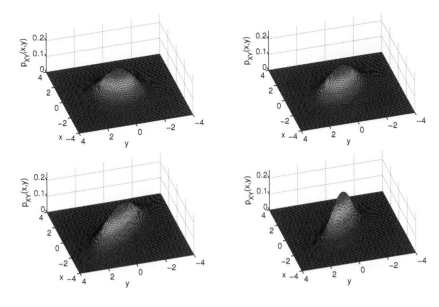

FIGURE 1.1
Bivariate normal distribution ($\mu_X = \mu_Y = 0$, $\sigma_X = \sigma_Y = 1$) with $\rho_{XY} = 0$, $\rho_{XY} = 0.25$, $\rho_{XY} = 0.5$, and $\rho_{XY} = 0.75$.

where μ_X, μ_Y locate the distribution and σ_X^2, σ_Y^2 quantify the spread in each direction. The parameter $0 \leq |\rho_{XY}| \leq 1$ is a shape parameter governing the degree to which the two random variables are correlated. To see why this is true, Figure 1.1 shows several bivariate normal distributions with $\mu_X = \mu_Y = 0$, $\sigma_X = \sigma_Y = 1$ for various ρ_{XY}. As $\rho \to 1$ the probability of finding Y in an interval $[y, y+dy]$ becomes completely specified by knowledge of $x \in [x, x+dx]$.

In many cases the joint probability model $p_{XY}(x, y)$ will be specified but we are more interested in knowing either $p_X(x)$ or $p_Y(y)$. To understand the relationship between the two, first consider the functions

$$P_{XY}(x, \infty) = Pr(-\infty < X \leq x, -\infty < Y \leq \infty) = P_X(x)$$
$$P_{XY}(\infty, y) = Pr(-\infty < X \leq \infty, \ -\infty < Y \leq y) = P_Y(y) \qquad (1.17)$$

where we have used the fact that the event consisting of all possible outcomes ($y \in [-\infty, \infty]$) occurs with probability 1 (Eqn. 1.2b). We may therefore write

$$p_X(x) = dP_X(x)/dx$$
$$= d/dx \int_{-\infty}^{x} \int_{-\infty}^{\infty} p_{XY}(u, v)\,du\,dv$$

$$= \int_{-\infty}^{\infty} p_{XY}(x, v) dv$$

$$p_Y(y) = \int_{-\infty}^{\infty} p_{XY}(u, y) du \tag{1.18}$$

which are the *marginal distributions* associated with the joint probability density. One can always form a marginal density by integrating over the support of the other random variables in the density function. Whether to call the distribution $p_X(x)$ a marginal distribution or just a distribution depends entirely on the ancestry of function and not on anything having to do with the properties of the function. When $p_X(x)$ is obtained by integration of a joint distribution over the other variable, it is called a "marginal distribution." Otherwise, it is called simply a "distribution." In the bivariate normal case (1.16) it is easy to show that the associated marginals are

$$p_X(x) = \frac{1}{\sqrt{2\pi\sigma_X^2}} e^{-\frac{1}{2\sigma_X^2}(x-\mu_X)^2}$$

$$p_Y(y) = \frac{1}{\sqrt{2\pi\sigma_Y^2}} e^{-\frac{1}{2\sigma_Y^2}(y-\mu_Y)^2} \tag{1.19}$$

i.e., each yields a normally distributed random variable.

In addition to joint probability models (probability of multiple events occurring) we can also form *conditional* probability models which yield the probability of an event occuring given knowledge of another event[2]. We begin by denoting the probability of the event $-\infty < X \le x$ given that we have already observed $-\infty < Y \le y$ as

$$P_X(x|y) = \frac{P_{XY}(x, y)}{P_Y(y)}. \tag{1.20}$$

Moreover, we can see that if the joint PDF can be factored as $p_{XY}(x, y) = p_X(x)p_Y(y)$ we have

$$p_X(x|y) = \frac{p_X(x)p_Y(y)}{p_Y(y)} = p_X(x) \tag{1.21}$$

which is simply another way of saying that knowledge of one variable has no bearing on the probability associated with the other. In fact, this result is used to define independence. Specifically, two random variables are said to be independent if their joint PDF (and CDF) factors

$$P_{XY}(x, y) = P_X(x)P_Y(y)$$

$$p_{XY}(x, y) = p_X(x)p_Y(y). \tag{1.22}$$

---

[2]Note that "knowledge" of another event does not necessarily mean that event has already occured, i.e., nowhere in the definition of conditional probability is a requirement that the events be temporally ordered.

Before concluding this section we mention that in some experiments we have a good probability model for a random variable X but will also be interested in developing a model for a function of that random variable, $Y = g(X)$. In this case we may use our knowledge of $p_X(x)$ and the functional form $g(\cdot)$ to derive $p_Y(y)$. First, consider the case where $g(\cdot)$ is a real, one-to-one, continuous function of its argument. For infinitesimal increments Δx, Δy, we may write $Pr(x < X \leq x + \Delta x) = Pr(y < Y \leq y + \Delta y)$, that is to say the probability of finding the random variable X in the interval $[x, x + \Delta x]$ is the same as that for Y on $[y, y + \Delta y]$. So in terms of the PDFs of X and Y we may write

$$p_Y(y)\Delta y \approx Pr(y < Y \leq y + \Delta y) = Pr(x < X \leq x + \Delta x) \approx p_X(x)\Delta x$$

so that

$$p_Y(y) = \lim_{\Delta x, \Delta_y \to 0} p_X(x)/|(\Delta y/\Delta x)|$$

$$= p_X\left(x = g^{-1}(y)\right) \left|\frac{dy\left(x = g^{-1}(y)\right)}{dx}\right|^{-1}$$

$$= p_X\left(g^{-1}(y)\right) \left|g'\left(g^{-1}(y)\right)\right|^{-1} \tag{1.23}$$

The absolute value in the denominator ensures that the result is non-negative as required of a PDF. Additionally, because the left-hand side is a function of the random variable Y, the right-hand side must be evaluated at $x = g^{-1}(y)$. Eqn. (1.23) says that for a one-to-one function, knowing both the derivative and the inverse are sufficient to construct the PDF for Y given the PDF for X. For other well-behaved continuous functions, e.g., $y = x^2$, the situation is only slightly more complicated. It can be shown that for continuous function that are not one-to-one, i.e., multiple values in the domain map to a single value in the range, Eqn. (1.23) becomes

$$p_Y(y) = \sum_{k=1}^{n_x} p_X\left(g_k^{-1}(y)\right) \left|g'\left(g_k^{-1}(y)\right)\right|^{-1} \tag{1.24}$$

where n_x is the number of values in the domain that map to a single value in the range and $x = g_k^{-1}(y)$ are the inverse mappings for each branch of the function. Finally, we may extend the above approach to vectors of random variables, e.g., $\mathbf{y} \equiv (y_1, y_2, \cdots, y_M)$, in which case Eqn. (1.23) becomes

$$p_Y(\mathbf{y}) = p_X\left(\mathbf{g}^{-1}(\mathbf{y})\right) |\mathbf{J}(\mathbf{g}^{-1}(\mathbf{y}))|^{-1} \tag{1.25}$$

where $\mathbf{J}$ is the Jacobian of the transformation, defined as the matrix

$$\mathbf{J} \equiv J_{ij} = \frac{\partial y_i}{\partial x_j}. \tag{1.26}$$

In later chapters we will use such transformations in the derivation of differential entropies for different probability density functions.

While there are many other aspects of probabilistic modeling, an understanding of the material presented here is sufficient to understand what follows. Our goal here was to establish a consistent notation for discussing such models, present some useful properties of the models, and describe some common manipulations (e.g., transformation of variables) that can aid in the simplification of probabilistic models.

1.2 Expectation and Moments

Say that we have modeled the outcome of an experiment x as a random variable X using the function $p_X(x)$. The probability of finding a particular outcome in the interval $[-\Delta_x, \Delta_x]$ is simply $\int_{-\Delta_x}^{\Delta_x} p_X(x)dx$. However, we might also be interested in "expected" or "likely" outcomes. In other words, if we repeat the experiment some number of times, what is the average value of X? How much variability do we expect in X? If we consider a function of X, what is its most likely output?

The answer to each of these questions requires us to define what we mean by "expected." For a probability distribution $p_X(x)$, define the expected value as

$$E[X] = \int_X xp_X(x)dx \tag{1.27}$$

where the integral is understood to be taken over the support of the random variable. Thus, it is appropriate to think of the expected value operator as a weighted average of the random variable X where the weightings are given by the values of the PDF. In other words, if we repeat the experiment many times, the average value of the outcomes would be close to the value given by $E[X]$. We might similarly define the expectation for other functions of X, e.g.,

$$E[X^2] = \int_X x^2 p_X(x)dx$$

$$E[X^3 - X^2 - X] = \int_X (x^3 - x^2 - x)p_X(x)dx$$

$$E[\sqrt{X}] = \int_X x^{1/2} p_X(x)dx \tag{1.28}$$

or, in general,

$$E[g(X)] = \int_X g(x)p_X(x)dx \tag{1.29}$$

where $g(X)$ is a deterministic function of the random variable. The same formalism can be extended to multiple random variables, e.g.,

$$E[XY] = \int_X \int_Y xy p_{XY}(x,y)\,dx\,dy$$

$$E[XYZ] = \int_X \int_Y \int_Z xyz p_{XYZ}(x,y,z)\,dx\,dy\,dz \qquad (1.30)$$

which are referred to as "joint expectations." In general, we have (for two random variables) $E[g(X,Y)] = \int_X \int_Y g(x,y) p_{XY}(x,y)\,dx\,dy$ for $g(X,Y)$ a deterministic function.

Certain expectations are more frequently used than others in describing experimental outcomes. We have already alluded to the *mean* of a random variable (the average) defined by Eqn. (1.27). Considering the Gaussian distribution (1.10) and carrying out the integral required of (1.27) yields $E[X] = \mu_X$. In fact the notation $\mu_X \equiv E[X]$ will be used throughout this book to denote the mean of a random variable, regardless of the probability model used. A second expectation that figures prominently in this work is the *variance* of a random variable

$$\sigma_X^2 \equiv E[(X - E[X])^2] \qquad (1.31)$$

Again, the notation σ_X^2 borrows from the Gaussian distribution where it can be seen that

$$\int_X (x - \mu_X)^2 \frac{1}{\sqrt{2\pi\sigma_X^2}} e^{-\frac{1}{2\sigma_X^2}(x-\mu_X)^2}\,dx = \sigma_X^2 \qquad (1.32)$$

The mean and variance are by far the most common descriptors of a random variable, however Eqn. (1.32) naturally extends to other *central moments* of a random variable, defined as

$$E[(X - E[X])^n] = \int_{-\infty}^{\infty} (x - \mu_X)^n p_X(x)\,dx. \qquad (1.33)$$

Now, for a zero-mean random variable the associated moments become simply $E[X^n] = \int_{-\infty}^{\infty} x^n p_X(x)\,dx$. In this case *all* moments of a random variable can be captured in a single function. Define the *characteristic* function as

$$\phi_X(it) = \int_{-\infty}^{\infty} e^{itx} p_X(x)\,dx \qquad (1.34)$$

which, by the definition of expectation (1.29), is $E[e^{itX}]$. Expanding this expectation as a power series gives

$$\phi_X(it) = E[e^{itx}] = 1 + it E[x] + \frac{-t^2}{2} E[x^2] + \frac{-it^3}{6} E[x^3] + \cdots + \frac{(it)^n}{n!} E[x^n] \qquad (1.35)$$

from which we see by inspection that

$$E[X^n] = \left. \frac{d^n \phi(it)}{dt^n} \right|_{t=0}. \tag{1.36}$$

The utility of defining such a function is not merely one of compact representation. The characteristic function is extremely useful in simplifying expressions that involve the transformation of random variables [29]. For this reason the characteristic function is provided later for each of the distributions presented in Chapter 4.

Not surprisingly, it is also common to define *joint* moments between random variables X and Y. Later, we will make reference to a particular joint moment, the *covariance*

$$\begin{aligned} C_{XY} &\equiv E[(X - E[X])(Y - E[Y])] \\ &= E\left[XY - E[X]Y - E[Y]X + E[X]E[Y]\right] \\ &= E[XY] - E[X]E[Y] \end{aligned} \tag{1.37}$$

This quantity is a frequently used descriptor of dependency among two random variables. In fact, for the bivariate normal distribution we have

$$\rho_{XY} = \frac{C_{XY}}{\sigma_X \sigma_Y} \tag{1.38}$$

i.e., the cross-correlation coefficient is simply the covariance, normalized by the product of the standard deviations of each random variable. As with many statistical properties, the normalized covariance is frequently denoted ρ_{XY}, even in situations where the joint probability model is non-Gaussian. The cross-correlation coefficient will be used extensively in Chapter 6.

1.3 Random Processes

We have just discussed joint distributions and covariance among two random variables. Most experiments, however, produce a temporal sequence of observations $\mathbf{x} \equiv (x(t_1), x(t_2), \cdots, x(t_N))$. For example, in many experiments a temporally varying signal $x(t)$ is sampled at discrete times $t_n = n\Delta_t$, $n = 1 \cdots N$ where Δ_t is the sampling interval. In this case the notation $x(n) \equiv x(n\Delta_t)$ is also sometimes used, i.e., the discrete time index is sufficient to define the sample time, provided the sampling interval is also specified. We will use both notations in this book, using continuous time for analytical treatment and discrete time in application. This choice will always be made explicit.

If the outcome of the experiment is uncertain it will be referred to as a *random process.* Moreover, each time the experiment is conducted, a different $N-$ sequence is produced and will henceforth be referred to as a *realization* of the experiment. If each observation in a given realization is modeled as a random variable, we have for the entire $N-$ sequence the random process model $\mathbf{X} \equiv (X(t_1), X(t_2), \cdots, X(t_N))$. Each random variable can be modeled with (possibly) a different probability model.

Just as with the $N = 2$ case, we would like to make probabilistic statements about joint events, for example the joint CDF for this random process is defined

$$P_{\mathbf{X}}(X(t_1) \leq x(t_1), X(t_2) \leq x(t_2), \cdots, X(t_N) \leq x(t_N)) \tag{1.39}$$

along with the joint PDF

$$p_{\mathbf{X}}(x(t_1), x(t_2), \cdots, x(t_N)) \tag{1.40}$$

where the latter can be expressed

$$p_{\mathbf{X}}(x(t_1), x(t_2), \cdots, x(t_N)) = \frac{\partial^N P_{\mathbf{X}}(X(t_1), X(t_2), \cdots, X(t_N))}{\partial x(t_1) \partial x(t_2) \cdots \partial x(t_N)} \tag{1.41}$$

in cases where the CDF is differentiable. Eqn. (1.40) is a model that quantifies the probability of having observed the particular sequence $x(t_1), x(t_2), \cdots x(t_N)$. Using this notation, analogous to Eqn. (1.18) we can define the marginal density associated with the nth random variable in the sequence

$$p_{X(t_n)}(x(t_n)) = \int_{\mathbb{R}^{N-1}} p_{X(t_1)X(t_2)\cdots X(t_N)}(x(t_1), x(t_2), \cdots, x(t_N)) dx(t_1) dx(t_2) \cdots$$
$$\times dx(t_{n-1}) dx(t_{n+1}) \cdots dx(t_N) \tag{1.42}$$

where the notation $\mathbb{R}^{N-1}$ denotes the multi-dimensional (infinite) integral over all variables other than $x(t_n)$.

We might also consider two different random processes $\mathbf{X}$ and $\mathbf{Y}$. The marginal density for the vector $\mathbf{X}$ given the joint density $p_{\mathbf{XY}}(\mathbf{x}, \mathbf{y})$ is simply

$$p_{\mathbf{X}}(\mathbf{x}) = \int_{-\infty}^{\infty} p_{\mathbf{XY}}(\mathbf{x}, \mathbf{y}) dy \tag{1.43}$$

where the integral is understood to be multi-dimensional, extending over each of the random variables in the vector $\mathbf{y}$. Likewise, statements of conditional probability and independence can be written

$$p_{\mathbf{X}}(\mathbf{x}|\mathbf{y}) = p_{\mathbf{XY}}(\mathbf{x}, \mathbf{y})/p_{\mathbf{Y}}(\mathbf{y}) \tag{1.44}$$

and

$$p_{\mathbf{XY}}(\mathbf{x}, \mathbf{y}) = p_{\mathbf{X}}(\mathbf{x})p_{\mathbf{Y}}(\mathbf{y}), \tag{1.45}$$

respectively. These latter properties will prove useful in Chapter 6 when we apply differential entropy to different engineering problems.

Just as we defined expectations of random variables we can define expectations of random processes. The only difference is that the random variable is now indexed by time. For example, we define the mean

$$\mu_X(t) \equiv E[X(t)] = \int_{-\infty}^{\infty} x(t) p_{\mathbf{X}}(x(t)) dx(t) \tag{1.46}$$

as the expected value of the random process at time t. If we were to repeat the experiment, collecting a different $x(t)$ each time (i.e., a different realization) $\mu_X(t)$ is interpreted as the average value of the random variable "X" recorded at time "t" (relative to the start of the experiment).

In general, to *completely* specify a discrete random process consisting of N observations one would have to specify all of the joint moments

$$E[X(t)]$$
$$E[X(t_1)X(t_2)]$$
$$E[X(t_1)X(t_2)X(t_3)]$$

$$\vdots$$

$$\tag{1.47}$$

In practice, however, it is far more common to focus only on the first few moments. For example, in this work it will be important to define the auto-correlation function

$$R_{XX}(t_n, t_m) \equiv E[X(t_n)X(t_m)] = \int_{\mathbb{R}^2} x(t_n)x(t_m) p_{\mathbf{X}}\left(x(t_n), x(t_m)\right) dx(t_n) dx(t_m).$$

$$\tag{1.48}$$

which, along with Eqn. (1.46) can be combined to form the *auto-covariance* function

$$C_{XX}(t_n, t_m) = E[(X(t_n) - \mu_X(t_n))(X(t_m) - \mu_X(t_m))]$$
$$= R_{XX}(t_n, t_m) - \mu_X(t_n)\mu_X(t_m). \tag{1.49}$$

The importance of (1.49) derives from the fact that it completely describes a large class of random processes. A common model for the observed sequence $\mathbf{x} = (x(t_1), x(t_2), \cdots, x(t_N))$ is the joint Gaussian PDF

$$p_{\mathbf{X}}(\mathbf{x}) = \frac{1}{2\pi^{N/2}|\mathbf{C_{XX}}|^{1/2}} e^{[-\frac{1}{2}(\mathbf{x}-\mu_{\mathbf{x}})^T \mathbf{C_{XX}}^{-1}(\mathbf{x}-\mu_{\mathbf{x}})]} \tag{1.50}$$

where

$C_{XX} =$

$$\begin{bmatrix} E[(X(t_1) - \mu_X(t_1))(X(t_1) - \mu_X(t_1))] & E[(X(t_1) - \mu_X(t_1))(X(t_2) - \mu_X(t_2))] & \cdots \\ E[(X(t_2) - \mu_X(t_2))(X(t_1) - \mu_X(t_1))] & E[(X(t_2) - \mu_X(t_2))(X(t_2) - \mu_X(t_2))] & \cdots \\ \vdots & \vdots & \ddots \\ E[(X(t_N) - \mu_X(t_N))(X(t_1) - \mu_X(t_1))] & E[(X(t_N) - \mu_X(t_N))(X(t_2) - \mu_X(t_2))] & \cdots \end{bmatrix}$$

$$\begin{matrix} \cdots & E[(X(t_1) - \mu_X(t_1))(X(t_N) - \mu_X(t_N))] \\ \cdots & E[(X(t_2) - \mu_X(t_2))(X(t_N) - \mu_X(t_N))] \\ \ddots & \vdots \\ \cdots & E[(X(t_N) - \mu_X(t_N))(X(t_N) - \mu_X(t_N))] \end{matrix} \qquad (1.51)$$

is the $N \times N$ covariance matrix associated with the observations and $|\cdot|$ takes the determinant. To see that this is just the N-dimensional analogue of (1.16) we may set $N = 2$ and recover (1.16) from (1.50). It can therefore be stated that the covariance matrix completely defines a joint Gaussian probability density function. Note that we could also define the covariance matrix between *two* random processes X and Y as

$$C_{XY}(t_n, t_m) = E\left[(X(t_n) - \mu_X(t_n))(Y(t_m) - \mu_Y(t_m))\right]. \qquad (1.52)$$

In the above development we have allowed each observation, i.e., each $x(t_n)$ the possibility of being modeled as a different random variable $X(t_n)$. Thus, each observation could follow a different CDF and PDF, e.g., $p_{X(t_1)}(x(t_1)) \neq p_{X(t_2)}(x(t_2))$. In fact, because the random variables are functions of time, the joint PDF associated with one realization could be entirely different from that associated with another. For many experiments, this level of flexibility is not required of the model. If we can probabilistically model the experimental outcome as being independent of time, that is each realization is modeled with the same joint probability distribution, we may write

$$p_{\mathbf{X}}(x(t_1), x(t_2), \cdots, x(t_N)) = p_{\mathbf{X}}(x(t_1 + \tau), x(t_2 + \tau), \cdots, x(t_N + \tau)) \quad (1.53)$$

for some constant time delay τ. Such an outcome is said to be *stationary* with respect to time. Invoking this assumption is a tremendous advantage, as the statistical properties just discussed are no longer time-dependent. For example, the mean of $X(t + \tau)$ is the same for all τ, i.e.,

$$E[X(t)] = E[X(t + \tau)] = \int_{-\infty}^{\infty} x p_{\mathbf{X}}(x) dx$$

$$= \mu_X. \qquad (1.54)$$

Properties like the auto-covariance also no longer depend on the absolute time index; however, they will depend on the relative temporal separation between the samples. The stationary auto-covariance matrix is given by

$$C_{XX}(t_n, t_m) = E[(X(t_n) - \mu_X(t_n))(X(t_m) - \mu_X(t_m))]$$

$$= E[(X(t_n) - \mu_X)(X(t_n + \tau) - \mu_X)$$

$$= C_{XX}(\tau) \qquad (1.55)$$

where in the last line we have simply noted that the joint distribution $p_{X(t_n)X(t_n+\tau)}(x(t_n), x(t_n+\tau))$ does not depend on t_n, but only on $\tau = t_m - t_n$. Thus, the stationary covariance matrix is not a two-dimensional function of t_n, t_m, but a one-dimensional function of the relative temporal separation between samples.

Under the assumption of stationarity other statistical properties will also depend only on a time lag τ. Obviously, the cross-correlation coefficient is one such property, defined as

$$\rho_{XY}(\tau) = C_{XY}(\tau)/\sigma_X\sigma_Y \tag{1.56}$$

where σ_X^2, σ_Y^2 are the stationary variances of the random processes X and Y, respectively. Like the stationary mean (1.54), these quantities are temporally constant, e.g., $\sigma_X^2 = E[(X(t_1) - \mu_X)(X(t_1) - \mu_X)] = C_{XX}(0)$. Equations (1.55) and (1.56) will be used extensively in Chapter 6 in demonstrating applications of differential entropy.

The above analyses extend to higher-order statistical properties as well. For example, assuming a stationary random process $\mathbf{X}$ we might define the third-order correlation function $C_{XXX}(\tau_1, \tau_2) = E[(X(t) - \mu_X)(X(t + \tau_1) - \mu_X)(X(t + \tau_2) - \mu_X)]$ as a function of the two delays τ_1, τ_2. These quantities are not studied in this work; however, they are of central importance to the field of higher-order spectral analysis and have many uses in the detection of non-linearity from time-series data [40].

As a final point, we mention a class of random process models based on conditional probability. Specifically, we define the $\mathcal{P}^{th}$ order Markov model associated with the random process $\mathbf{X}$ as a probabilistic model that obeys

$$p_{\mathbf{X}}(x(t_{n+1})|x(t_n), x(t_{n-1}), \cdots, x(t_{n-\mathcal{P}+1}), x(t_{n-\mathcal{P}}), \cdots)$$
$$= p_{\mathbf{X}}(x(t_{n+1})|x(t_n), x(t_{n-1}), \cdots, x(t_{n-\mathcal{P}+1})) \tag{1.57}$$

In other words, the probability of the random variable $X(t_{n+1})$ attaining the value $x(t_{n+1})$ is conditional on $X(t_n)$, $X(t_{n-1}), \cdots, X(t_{n-\mathcal{P}+1})$ attaining $x(t_n)$, $x(t_{n-1}), \cdots, x(t_{n-\mathcal{P}+1})$, but *not* on any values prior to this sequence. This is a generalized model for systems with finite memory, a specific class of which are the familiar auto-regressive models [20]. Markov models will be used later in Chapter 6 in defining the differential entropies associated with a measure of dynamical coupling.

1.4 Probability Summary

In this section we include a brief discussion of some aspects of probabilistic modeling that will be useful in our upcoming discussion of entropy. Returning to our axioms of probability (1.2b), the sum or integral of the probabilities over all possible outcomes must be unity, i.e.,

$$\sum_x f_X(x) = 1$$

$$\int_{-\infty}^{\infty} p_X(x)dx = 1$$

$$\sum_{x,y} f_{XY}(x,y) = 1$$

$$\int_{-\infty}^{\infty}\int_{-\infty}^{\infty} p_{XY}(x,y)dxdy = 1. \tag{1.58}$$

These are really statements of logical consistency that have been cast in the form of equations. Considering the discrete case, if we multiply the conditional probability $f_X(x|y)$ of observing x given that $Y = y$, by the probability $f_Y(y)$ that $Y = y$, then we must obtain the joint probability of observing *both* $X = x$ and $Y = y$. Furthermore, the joint probability cannot depend on which variable we choose to list first

$$f_{XY}(x,y) \equiv f_{YX}(y,x) \tag{1.59}$$

hence

$$f_X(x|y)f_Y(y) = f_{XY}(x,y) = f_{YX}(y,x) = f_Y(y|x)p_X(x). \tag{1.60}$$

This equation can be rearranged as follows

$$f_X(x|y) = \frac{f_Y(y|x)f_X(x)}{f_Y(y)}. \tag{1.61}$$

A similar argument holds for continuous random variables so that we may write

$$p_X(x|y) = \frac{p_Y(y|x)p_X(x)}{p_Y(y)}, \tag{1.62}$$

an expression referred to as *Bayes' theorem*.

The great utility of this result is that it tells us how to relate the two conditional probabilities $p_X(x|y)$ and $p_Y(y|x)$. For example, suppose we have obtained experimental data from a system that we believe is described by a certain physical model. Let x denote the experimental data and let y denote the model. Then $p_X(x|y)$ is the probability of observing the data x given that the model y is true, while $p_Y(y|x)$ is the probability that the model y is consistent with the observed data x. Presumably we know the details of our model so we can directly calculate $p_X(x|y)$. The individual PDF $p_Y(y)$ is referred to as the *prior* probability and must be specified *a priori*. The distribution $p_X(x)$ is a function of the data only and may be estimated.

We should point out that nothing demands that the two variables in the mixed probabilities $p_X(x|y)$ and $p_{XY}(x,y)$ be of the same type. For example, we can have a situation in which x is a discrete variable while y is a continuous variable. An important example of just this situation occurs in digital communications where the x's are the discrete alphabet of possible digital symbols sent by the transmitter and y corresponds to the analog, continuous noise voltage present at the receiver.

It is also worth noting that there are fundamentally different interpretations of probability itself. One school of thought views a probability distribution as the set of limiting values of a large number of trials or measurements (real or conceptual) made on a random variable. This so-called *frequentist* perspective therefore views a probability distribution as an intrinsic property of the process that gave rise to the variable.

In the second school are those who view the probability distribution as a model that predicts the outcome of any single measurement (real or conceptual) made on the variable [25]. The probability distribution therefore represents an anthropomorphic *state of knowledge* about the random variable. This is sometimes referred to as the *Bayesian* perspective as the entire concept of the prior PDF in Eqn. (1.61) or (1.62) requires such an interpretation. From this viewpoint, the distribution can certainly depend on the properties of the variable or the process but not solely on those properties. The distribution must also depend on the state of ignorance or uncertainty on the part of the observer. That is to say, it is the observer who *assigns* a probability distribution to the measurement of a random variable based on all the available information. From this perspective, every probability distribution or density is conditioned on all the information available at the time the distribution is assigned. In the literature this is often written as, for example, $p_X(x|I)$ instead of simply $p_X(x)$ to emphasize that all available information I was used in formulating $p_X(x)$. However, for reasons of compactness and clarity, the conditioning on I is usually dropped and we use simply $p_X(x)$.

To be sure, in specific instances it may be the case that the only means available for assigning probabilities is to make a large number of measurements, construct a histogram, and assign probabilities based on the frequencies of occurrence. However, based on our practical experience, the Bayesian viewpoint is the more useful one and leads to the fewest conceptual and calculational difficulties. In fact, at the end of Chapter 2 we will discuss a powerful application of the Bayesian approach in the form of the entropy maximization principle. This viewpoint has been clearly expounded and vigorously defended by Jaynes [18] and has been reinforced by Feynman and Hey [13].

Finally, we need to make one remark concerning the conditional probability density $p_X(x|y)$. Note that we have been careful to make X the subscript of this density because, even though this conditional density does depend on the value y, it is, in fact, a density only for variable X and must have units inverse to the units of x. Also, as a result of the y-dependence, we cannot assert any kind of normalization condition of the type in Eqn. (1.58). In fact, we can

TABLE 1.1
Units associated with commonly used probability density functions.

Probability Density Function	Units (where $[x]^{-1}$ indicates units inverse to the units of x)	Normalization
$p_X(x)$	$[x]^{-1}$	$\int_X p_X(x)dx = 1$
$p_Y(y)$	$[y]^{-1}$	$\int_Y p_Y(y)dy = 1$
$p_{XY}(x,y)$	$[x]^{-1}[y]^{-1}$	$\int_X \int_Y p_{XY}(x,y)dxdy = 1$
$p_X(x|y)$	$[x]^{-1}$	$\int_X p_X(x|y)dx \geq 0$

only assert that the integral is non-negative

$$\int_{-\infty}^{\infty} p_X(x|y)dx \geq 0 \tag{1.63}$$

See Table 1.1 for a description of the units associated with probability density functions.

2

The Concept of Entropy

The mathematical quantity we now call entropy arose from nearly independent lines of theoretical development in classical thermodynamics through the work of Clausius, statistical mechanics through the work of Boltzmann and Gibbs, and communications theory through the work of Shannon. At the end of the chapter we shall discuss in more detail these historical developments and the place entropy holds in these fields.

We will first consider entropy, joint entropy and conditional entropy corresponding to the different types of probability distributions discussed in the first chapter. From pedagogical, calculational and traditional points of view it is natural to define and discuss the entropies in exactly that order, and that is the order we shall use here. We will also find that the joint and conditional entropies retain their meaning with no conceptual difficulties in both the discrete and continuous cases.

Through the years the language associated with entropy, especially within the information sciences, has seen a dichotomous development. In one approach, entropy is said to be a measure of uncertainty while, in the other camp, entropy is said to be a measure of information. The distinction may initially seem trivial since it could be argued that information and uncertainty are merely two sides of the same coin. However, in our practical experience, we have found the interpretation of entropy as uncertainty to be much more useful and much less prone to misunderstanding. The larger the uncertainty about X, the larger the entropy associated with the probability distribution $p_X(x)$ of X. Hence, the entropy tells us something about the shape of the distribution function. Given two probability distributions defined over the same range, the broader, flatter distribution will have the larger entropy, a fact that is consistent with our defining probability as a model for uncertainty. For example, we are much less certain about the outcome of measurement modeled by a broad, flat distribution than we are about one governed by a distribution having a single sharp peak.

We begin the chapter with definitions, discussion and examples of entropy for both discrete and continuous variables. At the end of the chapter we will delve more deeply into the distinctions in meaning for discrete and differential entropies and we will discuss the importance of the concept of entropy in two foundational areas of modern science: statistical mechanics and communication theory.

2.1   Discrete Entropy

The concept of entropy is usually formulated first for discrete random variables; that is, X is a discrete random variable with potential states $\{x_1, x_2, \cdots, x_m, \cdots\}$ and probability mass function (PMF) $f_X(x_m) \equiv Pr(X = x_m)$. Using this notation, the discrete entropy is expressed as

$$H_X = \sum_m f_X(x_m) \log\left[\frac{1}{f_X(x_m)}\right]$$
$$= -\sum_m f_X(x_m) \log\left[f_X(x_m)\right]. \tag{2.1}$$

This log is usually taken to be $\log_2$ in which case the entropy is given in units of "bits." If the logarithm is taken in the base e, then the entropy is given in units of "nats." By convention, we assume that if $f_X(x_m) = 0$ for any m, that $0\log_2(0) = 0$. It is clear from the first formulation above that $H_X \geq 0$ and that $H_X = 0$ only in the degenerate case where X takes on only a single value with probability 1.

Example 1

Suppose we make four tosses of an honest coin and keep track of the succession of heads and tails. The outcome of each toss (head or tail) is modeled as a random variable X with a binomial probability mass function $f_X(x_m) = p^{x_m}(1-p)^{1-x_m}$ where p is the probability of getting a head on a single trial. If all four trials are assumed independent and we assume a fair coin, i.e., $p = 0.5$, the probability assigned to the vector outcome of the four trials is uniform, with each of the $2^4 = 16$ equally likely outcomes assigned probability $f_X(x_m) = \frac{1}{16}$, $m = 1 \cdots 16$. The entropy of this random process is therefore

$$H_X = \sum_{i=1}^{16} \frac{1}{16} \log_2(16) = 4 \text{ bits}$$

which is indeed the number of bits needed to describe X since all outcomes can be denoted by binary strings of the same length (four) and we have no a priori means of condensing this information further. If there are 2^n equally likely outcomes, $H_X = n$ bits, and the entropy is monotonically increasing as n increases. This result is consistent with our intuitive notion of entropy in that as the number of possible states increases so too does our uncertainty in the outcome of the experiment modeled by $f_X(x_m)$. ∎

Example 2

The Washington Nationals baseball team has an attraction called the "Presidents race" where, during the fourth inning, massively headed costumed caricatures of four U.S. presidents — Abe Lincoln, George Washington,

Thomas Jefferson, and Teddy Roosevelt — race each other around the base-ball field. Abe has proven to be the best racer, George and Thomas are about even, and Teddy hardly ever wins[1]. Let's set the probabilities of winning at $\{\frac{1}{2}, \frac{1}{4}, \frac{15}{64}, \frac{1}{64}\}$ so the distribution is non-uniform. The discrete entropy associated with the outcome of the race would be

$$H_X = \frac{1}{2}\log_2(2) + \frac{1}{4}\log_2(4) + \frac{15}{64}\left(\log_2(64) - \log_2(15)\right) + \frac{1}{64}\log_2(64)$$
$$= \frac{1}{2} + \frac{1}{2} + \frac{15}{64}(6 - 3.907) + \frac{6}{64} = 1.58 \text{ bits}$$

as compared to the uniform case where the entropy would be 2 bits. Now suppose you are attending the game and you want to send a text message to your spouse at home telling who won the Presidents Race. You could have indexed the four presidents ahead of time and texted a number from 0 to 3; this will require 2 bits for any result. But it would make sense to use short messages for the more likely winners and longer ones for unlikely winners. For example, you could use the bit strings: 0, 10, 110, 111, for Abe, George, Tom, and Teddy in that order. The average number of bits sent in this case would be 1.75, which would be a significant reduction. This example illustrates the fact that the entropy represents the inherent average number of bits needed to describe a random variable; it provides a lower bound on the average number of bits needed for transmission of its values. Discrete distributions can therefore be directly compared based on their fundamental degree of uncertainty, as expressed by their (discrete) entropy value. ■

Recalling expression (1.29), we see that the (discrete) entropy can also be expressed as an expected value

$$H_X = E\left[\log_2\left(\frac{1}{f_X(x_m)}\right)\right]. \tag{2.2}$$

For a pair of discrete random variables $X : \{x_1, x_2, \cdots, x_m, \cdots\}$ and $Y : \{y_1, y_2, \cdots, y_n, \cdots\}$, the joint entropy can be similarly defined as

$$H_{XY} = E\left[\log_2\left(\frac{1}{f_{XY}(x_m, y_n)}\right)\right]$$
$$= \sum_m \sum_n f_{XY}(x_m, y_n)\log_2\left[\frac{1}{f_{XY}(x_m, y_n)}\right] \tag{2.3}$$

where $f_{XY}(x_m, y_n)$ is the joint PMF associated with X and Y.

---

[1]Teddy had actually recorded 524 consecutive losses until his first victory on 10/3/12. He followed this up with three consecutive victories bringing his winning percentage at the time of this writing to roughly half that used in this example.

We may also define the conditional entropy

$$H_{Y|X} = E\left[\log_2\left(\frac{1}{f_Y(y_n|x_m)}\right)\right]$$

$$= \sum_m \sum_n f_{XY}(x_m, y_n)\log_2\left(\frac{1}{f_Y(y_n|x_m)}\right)$$

$$= \sum_m f_X(x_m)\sum_n f_Y(y_n|x_m)\log_2\left(\frac{1}{f_Y(y_n|x_m)}\right)$$

$$= \sum_m f_X(x_m)H_{Y|X=x_m} \tag{2.4}$$

which relies on the conditional probability distributions discussed in Section (1.1). In general we see that $H_{Y|X} \neq H_{X|Y}$. However, the following useful properties are valid

$$H_{XY} = H_X + H_{Y|X} \text{ (chain rule)} \tag{2.5a}$$

$$H_{X,Y|Z} = H_{X|Z} + H_{Y|X,Z} \tag{2.5b}$$

$$H_X - H_{X|Y} = H_Y - H_{Y|X} \tag{2.5c}$$

$$H_{X|Y} \leq H_X \tag{2.5d}$$

with equality holding in (2.5d) if and only if X and Y are statistically independent.

Let us now examine in more detail what the properties (2.5a-2.5d) mean. The chain rule (2.5a) says that the joint uncertainty in X and Y is the sum of the uncertainty in X alone plus the uncertainty in Y given that X is known. Now $H_{Y|X} \leq H_Y$ since the uncertainty in Y cannot increase once the outcome of X is known (2.5d). Hence $H_{XY} \leq H_X + H_Y$, that is, the joint uncertainty in X and Y cannot be greater than the individual uncertainties. The second expression (2.5b) is an extension of the chain rule for the case where the joint probability for X and Y is now conditioned on knowledge of the outcome of a third variable Z. The result (2.5c) is a remarkable result. Recall that H_X and H_Y are measures of the uncertainty in predicting the outcomes of X and Y individually while $H_{X|Y}$ and $H_{Y|X}$ are measures of the uncertainties given that the outcome of the other variable is known. In general, for arbitrary random variables X and Y, $H_X \neq H_Y$ and $H_{X|Y} \neq H_{Y|X}$. However, Eqn. (2.5c) says that the reduction in uncertainty in X arising from knowledge of the outcome of Y, that is $H_X - H_{X|Y}$, is exactly equal to the reduction in uncertainty in Y due to knowledge of the outcome of X. And that this is true regardless of the relationship (if any) between X and Y! This quantity has such importance that it is given the name "mutual information" and it is discussed in some detail below. Given our encouragement for the interpretation of entropy as average uncertainty, the term "mutual information" is thus unfortunate.

TABLE 2.1

Joint probability mass function $f_{XY}(x_m, y_n)$ for Example 3.

$Y \backslash X$	x_1	x_2	x_3	x_4
y_1	$\frac{1}{4}$	$\frac{1}{32}$	$\frac{1}{16}$	$\frac{1}{4}$
y_2	$\frac{1}{8}$	$\frac{1}{8}$	$\frac{1}{32}$	0
y_3	$\frac{1}{16}$	$\frac{1}{32}$	$\frac{1}{32}$	0

Perhaps "joint reduction in uncertainty" would have been more accurate, but the language is unlikely to change at this point.

Example 3

Suppose X and Y are discrete random variables with joint distribution given in Table 2.1. Adding the rows gives the marginal distribution of Y which has probability values $\{\frac{19}{32}, \frac{9}{32}, \frac{1}{8}\}$; adding the columns yields the marginal distribution of X with probability values $\{\frac{7}{16}, \frac{3}{16}, \frac{1}{8}, \frac{1}{4}\}$. Consequently,

$$H_X = \frac{7}{16} \log_2 \left(\frac{16}{7}\right) + \frac{3}{16} \log_2 \left(\frac{16}{3}\right) + \frac{1}{8} \log_2 (8) + \frac{1}{4} \log_2 (4) = 1.85 \text{ bits}$$

$$H_Y = \frac{19}{32} \log_2 \left(\frac{32}{19}\right) + \frac{9}{32} \log_2 \left(\frac{32}{9}\right) + \frac{1}{8} \log_2 (8) = 1.34 \text{ bits}$$

and

$$H_{X|Y} = \sum_{n=1}^{3} Pr(Y = y_n) H_{X|Y=y_n}$$

$$= \frac{19}{32} H \left\{\frac{8}{19}, \frac{1}{19}, \frac{2}{19}, \frac{8}{19}\right\} + \frac{9}{32} H \left\{\frac{4}{9}, \frac{4}{9}, \frac{1}{9}, 0\right\} + \frac{1}{8} H \left\{\frac{1}{2}, \frac{1}{4}, \frac{1}{4}, 0\right\}$$

$$= \frac{19}{32} \left[\frac{8}{19} \log_2 \left(\frac{19}{8}\right) + \frac{1}{19} \log_2 (19) + \frac{2}{19} \log_2 \left(\frac{19}{2}\right) + \frac{8}{19} \log_2 \left(\frac{19}{8}\right)\right]$$

$$+ \frac{9}{32} \left[\frac{4}{9} \log_2 \left(\frac{9}{4}\right) + \frac{4}{9} \log_2 \left(\frac{9}{4}\right) + \frac{1}{9} \log_2 (9) + 0\right]$$

$$+ \frac{1}{8} \left[\frac{1}{2} \log_2 (2) + \frac{1}{4} \log_2 (4) + \frac{1}{4} \log_2 (4) + 0\right] = 1.54 \text{ bits}$$

$$H_{Y|X} = \sum_{m=1}^{4} Pr(X = x_m) H_{Y|X=x_m}$$

$$= \frac{7}{16} H \left\{\frac{4}{7}, \frac{2}{7}, \frac{1}{7}\right\} + \frac{3}{16} H \left\{\frac{1}{6}, \frac{2}{3}, \frac{1}{6}\right\} + \frac{1}{8} H \left\{\frac{1}{2}, \frac{1}{4}, \frac{1}{4}\right\} + \frac{1}{4} H \left\{1, 0, 0\right\}$$

$$= \frac{7}{16}\left[\frac{4}{7}\log_2\left(\frac{7}{4}\right) + \frac{2}{7}\log_2\left(\frac{7}{2}\right) + \frac{1}{7}\log_2\left(7\right) \right]$$

$$+ \frac{3}{16}\left[\frac{1}{6}\log_2\left(6\right) + \frac{2}{3}\log_2\left(\frac{3}{2}\right) + \frac{1}{6}\log_2\left(6\right) \right]$$

$$+ \frac{1}{8}\left[\frac{1}{2}\log_2\left(2\right) + \frac{1}{4}\log_2\left(4\right) + \frac{1}{4}\log_2\left(4\right) \right] + \frac{1}{4}(0) = 1.03 \text{ bits}$$

$$H_{XY} = H_X + H_{Y|X}$$
$$= H_Y + H_{X|Y}$$
$$= 2.88 \text{ bits}$$

Finally,

$$H_X - H_{X|Y} = 1.85 - 1.54 = 0.31$$
$$H_Y - H_{Y|X} = 1.34 - 1.03 = 0.31. \qquad (2.6)$$

■

Let $X : \{x_1, x_2, \cdots x_m, \cdots\}$ and $Y : \{y_1, y_2, \cdots, y_n, \cdots\}$ be discrete random variables modeled with joint PMF $f_{XY}(x_m, y_n)$ and marginal densities $f_X(x_m)$ and $f_Y(y_n)$, respectively. The *mutual information* can then be expressed as

$$I_{XY} = E\left[\log_2\left(\frac{f_{XY}(x_m, y_n)}{f_X(x_m)f_Y(y_n)} \right) \right]$$
$$= \sum_m \sum_n f_{XY}(x_m, y_n)\log_2\left(\frac{f_{XY}(x_m, y_n)}{f_X(x_m)f_Y(y_n)} \right). \qquad (2.7)$$

This particular function quantifies the difference between the joint probability model $f_{XY}(x_m, y_n)$ and the product of the individual models $f_X(x_m)f_Y(y_n)$. We recognize that $f_{XY}(x_m, y_n) = f_X(x_m)f_Y(y_m)$ if X and Y are statistically independent. Thus, the mutual information quantifies how close two random variables are to being independent. The mutual information is always greater than or equal to zero, with equality if and only if X and Y are statistically independent.

Mutual information and entropy have the following relationships:

$$I_{XY} = H_X - H_{X|Y} \qquad (2.8a)$$

$$I_{XY} = H_Y - H_{Y|X} \qquad (2.8b)$$

$$I_{XY} = I_{YX} \qquad (2.8c)$$

$$I_{XX} = H_X \qquad (2.8d)$$

$$I_{XY} = H_X + H_Y - H_{XY} \qquad (2.8e)$$

Example 4

For the discrete random variables X and Y of Example 3, the mutual information is given by

$$I_{XY} = H_X - H_{X|Y} = H_Y - H_{Y|X} = 0.31 \text{ bits}$$

■

Example 5

Suppose two professional baseball teams, the Astros (A) and Blue Jays (B), are playing each other in the World Series and that they are evenly matched (in each game). The World Series is a best-of-seven competition where the Series ends when one team is the first to win 4 games. There are two ways the Series can end in just 4 games

$$AAAA \quad - \quad \text{The Astros win 4 straight}$$
$$BBBB \quad - \quad \text{The Blue Jays win 4 straight}$$

Each of these occurs with probability $\frac{1}{2^4}$. There are 8 possibilities for a 5-game Series:

$$
\begin{array}{cc}
AAABA & BBBAB \\
AABAA & BBABB \\
ABAAA & BABBB \\
BAAAA & ABBBB
\end{array}
$$

Since the winning team always wins the last game, the losing team in this case must win one of the first 4 games, so there are $2\binom{4}{1} = 8$ possibilities, each occurring with probability $\frac{1}{2^5}$. Likewise, there are $2\binom{5}{2} = 20$ possibilities for a 6-game Series, each with probability $\frac{1}{2^6}$, and $2\binom{6}{3} = 40$ possibilities for a 7-game Series, each with probability $\frac{1}{2^7}$.

Let X be the discrete random variable representing the 70 possible outcomes of the Series and Y be the number of games played. The joint distribution table is given in Table 2.2.

The marginal distribution of Y has probability values $\{\frac{1}{8}, \frac{1}{4}, \frac{5}{16}, \frac{5}{16}\}$ and the marginal distribution of X has probability values

$$\left\{ \frac{1}{16}, \frac{1}{16}, \frac{1}{32}, \cdots, \frac{1}{32} \text{ (8 times)}, \frac{1}{64}, \cdots, \frac{1}{64} \text{ (20 times)}, \frac{1}{128}, \cdots, \frac{1}{128} \text{ (40 times)} \right\}.$$

Then we can compute the discrete entropies for each distribution:

$$H_X = 2\left[\frac{1}{16} \log_2 (16) \right] + 8\left[\frac{1}{32} \log_2 (32) \right] + 20\left[\frac{1}{64} \log_2 (64) \right]$$
$$+ 40\left[\frac{1}{128} \log_2 (128) \right]$$

TABLE 2.2
Joint probability mass function $f_{XY}(x_m, y_n)$ for the possible outcomes of the
World Series (X) and the number of games played (Y).

$Y \backslash X$	AAAA	BBBB	8 (5 game outcomes)	20 (6 game outcomes)	40 (7 game outcomes)
4	$\frac{1}{16}$	$\frac{1}{16}$	0	0	0
5	0	0	$\overbrace{\frac{1}{32} \cdots \frac{1}{32}}^{\text{8 of these}}$	0	0
6	0	0	0	$\overbrace{\frac{1}{64} \cdots \frac{1}{64}}^{\text{20 of these}}$	0
7	0	0	0	0	$\overbrace{\frac{1}{128} \cdots \frac{1}{128}}^{\text{40 of these}}$

$$= 2\left(\frac{4}{16}\right) + 8\left(\frac{5}{32}\right) + 20\left(\frac{6}{64}\right) + 40\left(\frac{7}{128}\right) = 5.81 \text{ bits}$$

$$H_Y = \frac{1}{8}\log_2(8) + \frac{1}{4}\log_2(4) + \frac{5}{16}\log_2\left(\frac{16}{5}\right) + \frac{5}{16}\log_2\left(\frac{16}{5}\right)$$

$$= 1.92 \text{ bits}$$

and the conditional entropies

$$H_{X|Y} = \sum_{n=1}^{4} Pr(Y = y_n) H_{X|Y=y_n}$$

$$= \frac{1}{8}H\left\{\frac{1}{2}, \frac{1}{2}\right\} + \frac{1}{4}H\left\{\frac{1}{8}, \cdots, \frac{1}{8} \text{ (8 of these)}\right\}$$

$$+ \frac{5}{16}H\left\{\frac{1}{20}, \cdots, \frac{1}{20} \text{ (20 of these)}\right\} + \frac{5}{16}H\left\{\frac{1}{40}, \cdots, \frac{1}{40} \text{ (40 of these)}\right\}$$

$$= \frac{1}{8}\left[2\left(\frac{1}{2}\log_2(2)\right)\right] + \frac{1}{4}\left[8\left(\frac{1}{8}\log_2(8)\right)\right] + \frac{5}{16}\left[20\left(\frac{1}{20}\log_2(20)\right)\right]$$

$$+ \frac{5}{16}\left[40\left(\frac{1}{40}\log_2(40)\right)\right]$$

$$= 3.89 \text{ bits}$$

$$H_{Y|X} = \sum_{m=1}^{70} Pr(X = x_m) H_{Y|X=x_m}$$

But here Y is completely determined when X is set to be a particular Series outcome so each of the $H_{Y|X=x_m}$ is zero and therefore

$$H_{Y|X} = 0.$$

The joint entropy is then given by $H_{XY} = H_X + H_{Y|X} = H_Y + H_{X|Y} = 5.81$ bits and the mutual information is given by

$$I_{XY} = H_X - H_{X|Y} = H_Y - H_{Y|X} = 1.92 \text{ bits}$$

■

Discrete entropy will be calculated for some important discrete distributions in Chapter 3. However, before concluding this section, it should be emphasized that the uniform distribution, in which each outcome is equally likely, holds a special place in discrete entropy. For N equally likely discrete outcomes, that is when $f_X(x_m) = 1/N$ for all m, we have seen that the entropy is $H_X = \log_2(N)$ bits. In fact, this is the maximum possible entropy for a probability mass function over N states. (A proof using Lagrange multipliers is given in Appendix 7.1).

2.2 Differential Entropy

The concept of entropy for continuous distributions was also presented in Shannon's original paper [48] and is referred to as the *differential entropy*. For a continuous random variable X with probability density function $p_X(x)$, the differential entropy is given by

$$h_X = -\int_S p_X(x) \log\left(p_X(x)\right) dx \tag{2.9}$$

where $S = \{x | p_X(x) > 0\}$ is the support set of X. The log function is again taken to be $\log_2$ giving the entropy in bits. As in the discrete case, if the log is taken in the base e, then the entropy is given units "nats."[2] Differential entropy retains many of the properties of its discrete counterpart, but with some important differences. Chief among these is the fact that differential entropy may take on any value between $-\infty$ and ∞ (recall discrete entropy is always non-negative). For example, we will show that in the limiting case where the density function is the Dirac Delta function the differential entropy is, in fact, $h_X = -\infty$.

Differential entropy is not a number, as in discrete entropy, but rather a function of one or more parameters that describe the associated probability

---

[2]The formula for the differential entropy certainly represents the natural extension to the continuous case, but this transition from discrete to continuous random variables must be handled carefully as we shall see.

distribution. Differential entropy represents not an absolute measure of uncertainty, but rather measures relative uncertainty or changes in uncertainty. To compare the entropy of two continuous distributions, one must choose a common statistic by which to evaluate the entropy functions. This is done in Chapter 5 based upon the variance statistic.

Let us examine how "natural" is the definition of differential entropy for a continuous random variable X. Assume that the probability density function $p_X(x)$ is Riemann integrable and thus continuous. Let us divide S into bins, each of length Δ. By the mean value theorem, within each bin $[n\Delta, (n+1)\Delta]$ there is a number x_n^* such that

$$\int\limits_{n\Delta}^{(n+1)\Delta} p_X(x)dx = p_X(x_n^*)\Delta.$$

Then consider the discrete random variable X_Δ obtained by consolidating all the x's in the bin $[n\Delta, (n+1)\Delta]$ to the point x_n^*, with probabilities $P_n = p_X(x_n^*)\Delta$. The discrete entropy of this discrete variable is then given by

$$\begin{aligned}
h_{X_\Delta} &= -\sum_n P_n \log(P_n) \\
&= -\sum_n p_X(x_n^*)\Delta \log\left(p_X(x_n^*)\Delta\right) \\
&= -\sum_n p_X(x_n^*)\Delta \left[\log\left(p_X(x_n^*)\right) + \log(\Delta)\right] \\
&= -\sum_n p_X(x_n^*) \log\left(p_X(x_n^*)\right)\Delta - \log(\Delta)
\end{aligned}$$

where we have used $\sum_n p_X(x_n^*)\Delta = \sum_n P_n = \int_S p_X(x)dx = 1$. By the definition of the Riemann integral, the first term approaches the differential entropy of X as $\Delta \to 0$. But there is the extra term $-\log(\Delta)$ which approaches ∞ as $\Delta \to 0$. So the difference between the differential entropy of a continuous random variable and the discrete entropy of its discretization is

$$h_X - h_{X_\Delta} = \log(\Delta)$$

which approaches $-\infty$ as $\Delta \to 0$. For consistency, we therefore define the differential entropy of any discrete random variable to be $-\infty$. It is because of this extra term that we can only make relative comparisons of differential entropies.

The above discussion is relevant to how one treats the entropy of a set of values obtained in a physical experiment. Suppose $\{x(1), x(2), \cdots, x(N)\}$ is a set of independent samples from a continuous random process $\mathbf{X}(t)$ whose underlying probability density function is $p_{\mathbf{X}}(\mathbf{x})$. If $p_{\mathbf{X}}(\mathbf{x})$ is known, then one calculates the differential entropy h_X. It is known that $Pr\{x(1), x(2), \cdots, x(N)\}$

is very near to 2^{-Nh_X} with high probability, that is,

$$-\frac{1}{N} \log_2 \left(p_{\mathbf{X}}(x(1), x(2), \cdots, x(N)) \right) \text{ approaches } h_X.$$

This has a geometrical interpretation in N−space $\mathbb{R}^N$. If A_ϵ is the set of all typical sequences in $\mathbb{R}^N$, that is, the set of all sequences for which the above expression is within ϵ of h_X, then this set has a volume of approximately 2^{Nh_X}. So low entropy corresponds to the sample values being contained within a small volume in N−space; high entropy corresponds to sample values with a large dispersion.

But how should one proceed if the underlying probability density function of X is not known? There seem to be two ways forward; both of which depend on how well the sample represents the underlying distribution; that is, whether the sample is typical. One approach is to consider the underlying distribution to be discrete; then the sample values are placed into bins and the fraction of samples in each bin is used to assign a probability. In this case, the discrete entropy will vary depending on the binning, so we need to have a natural bin size, say equal to measurement precision, and a large sample size. The second appproach is to form the sample data into a histogram and fit this histogram with a continuous probability distribution and calculate the corresponding differential entropy. Again, the validity of this approach will depend on how typical the data sequence is and how good is the fit of the distribution. Two words of caution: First, the values of entropy obtained by these two approaches will be different due to the $\log(\Delta)$ term. Secondly, in the first approach the discrete entropy will be a pure number; in the second approach the differential entropy may have units (see the note following Table 2.3).

For continuous random variables X and Y, joint differential entropy, conditional differential entropy, and mutual information are defined in a way analogous to their discrete counterparts

$$h_{XY} = \int \int p_{XY}(x, y) \log_2 \left(\frac{1}{p_{XY}(x, y)} \right) dx dy \tag{2.10}$$

$$h_{Y|X} = \int \int p_{XY}(x, y) \log_2 \left(\frac{1}{p_Y(y|x)} \right) dx dy \tag{2.11}$$

$$I_{XY} = \int \int p_{XY}(x, y) \log_2 \left(\frac{p_{XY}(x, y)}{p_X(x) p_Y(y)} \right) dx dy. \tag{2.12}$$

All of the properties (2.5a) through (2.5d), (2.8a) through (2.8c), and (2.8e) carry over to the continuous case. Also the relative entropy and the mutual information are greater than or equal to 0 as in the discrete case. Property (2.8d) cannot be valid in the continuous case since I_{XX} is never negative but h_X may be. Other properties include:

$$h_{X+k} = h_X \text{ where } k \text{ is a constant} \tag{2.13}$$

$$h_{kX} = h_X + \log_2(|k|) \text{ where } k \text{ is a constant} \tag{2.14}$$

If $Y = f(X)$ is a differentiable function of X, then

$$h_Y \leq h_X + E\left[\log_2\left(\left|\frac{df}{dx}\right|\right)\right] \tag{2.15}$$

with equality if and only if f has an inverse.

In the following chapter we derive the differential entropy for a large number of continuous random variables. For example, as given in Table 2.3, the differential entropy for the Normal distribution is

$$h_X = \frac{1}{2}\log_2\left(2\pi e\sigma^2\right) \tag{2.16}$$

In fact, if X is any continuous random variable with variance σ^2, then its entropy is less than or equal to that of the Normal entropy with equality only for the Normal distribution itself. This relationship will be illustrated in Chapter 5 and will be proven using Lagrange multipliers in Appendix 7.1.

Example 7 Mutual Information for Jointly Normal Random Variables

Consider two jointly Normal random variables X and Y with joint PDF

$$p_{XY}(x,y) = \frac{1}{2\pi\sigma_X\sigma_Y\sqrt{1-\rho_{XY}^2}}e^{-\frac{1}{2}(1-\rho_{XY}^2)\left[\frac{(x-\mu_Y)^2}{\sigma_X^2} + \frac{(y-\mu_Y)^2}{\sigma_Y^2} - \frac{2\rho_{XY}(x-\mu_X)(y-\mu_Y)}{\sigma_X\sigma_Y}\right]} \tag{2.17}$$

This probability model was introduced in Chapter 1 where it was further shown that the associated marginal distributions followed a Normal PDF with means μ_X, μ_Y and variances σ_X^2, σ_Y^2, respectively. From Eqn. (2.16) we have that the individual entropies associated with X and Y are

$$h_X = \frac{1}{2}\log_2\left(2\pi e\sigma_X^2\right)$$

$$h_Y = \frac{1}{2}\log_2\left(2\pi e\sigma_Y^2\right) \tag{2.18}$$

respectively. The joint entropy is given by the integral

$$h_{XY} = -\int_X\int_Y p_{XY}(x,y)\log_2\left(p_{XY}(x,y)\right)dxdy$$

$$= -E[\log_2\left(p_{XY}(x,y)\right)]$$

$$= \log_2\left(2\pi\sigma_X\sigma_Y\sqrt{1-\rho_{XY}^2}\right) + \frac{1}{2(1-\rho_{XY}^2)}E\left[\frac{(x-\mu_x)^2}{\sigma_x^2}\right.$$

$$\left. + \frac{(y-\mu_y)^2}{\sigma_y^2} - \frac{2\rho_{XY}(x-\mu_x)(y-\mu_y)}{\sigma_x\sigma_y}\right] \times \frac{1}{\ln(2)}$$

$$= \log_2\left(2\pi\sigma_X\sigma_Y\sqrt{1-\rho_{XY}^2}\right) + \frac{1}{2(1-\rho_{XY}^2)}\left[1 + 1 - 2\rho_{XY}^2\right] \times \frac{1}{\ln(2)}$$

$$= \log_2\left(2\pi e\sigma_X\sigma_Y\sqrt{1-\rho_{XY}^2}\right) \tag{2.19}$$

where we have used the identity $\log_2(e) = 1/\ln(2)$. Thus, the mutual information becomes

$$I_{XY} = h_X + h_Y - h_{XY}$$

$$= \frac{1}{2}\log_2\left(4\pi^2 e^2 \sigma_X^2 \sigma_Y^2\right) - \log_2\left(2\pi e \sigma_X \sigma_Y \sqrt{1 - \rho_{XY}^2}\right)$$

$$= \frac{1}{2}\log_2\left(4\pi^2 e^2 \sigma_X^2 \sigma_Y^2\right) - \log_2\left(4\pi^2 e^2 \sigma_X^2 \sigma_Y^2 (1 - \rho_{XY}^2)\right)^{1/2}$$

$$= \frac{1}{2}\log_2\left(\frac{4\pi^2 e^2 \sigma_X^2 \sigma_Y^2}{4\pi^2 e^2 \sigma_X^2 \sigma_Y^2 (1 - \rho_{XY}^2)}\right)$$

$$= -\frac{1}{2}\log_2\left(1 - \rho_{XY}^2\right) \tag{2.20}$$

and is a function of the cross-correlation coefficient only. Note that when $\rho_{XY} = 0$, that is, when X, Y are independent, $I_{XY} = 0$ and there is no reduction in uncertainty about X when Y is known. Also as $|\rho_{XY}| \to 1$, $I_{XY} \to \infty$. It is perhaps not surprising that the amount of information common to two jointly Gaussian random variables is dictated solely by ρ_{XY}. This is an important relationship in the study of differential entropy and will be generalized further in Chapter 6. ∎

Example 8

Consider the joint probability density function

$$p_{XY}(x,y) = \begin{cases} 1 - \frac{3}{4}x - \frac{1}{8}y & \text{for } 0 < x < 1,\ 0 < y < 2 \\ 0 & \text{otherwise} \end{cases}$$

The marginal distributions are

$$p_X(x) = \int_0^2 (1 - \frac{3}{4}x - \frac{1}{8}y)dy = y - \frac{3}{4}xy - \frac{1}{8}\frac{y^2}{2}\Big|_0^2 = \frac{7}{4} - \frac{3}{2}x \text{ for } 0 < x < 1$$

$$p_Y(y) = \int_0^1 (1 - \frac{3}{4}x - \frac{1}{8}y)dx = x - \frac{3}{4}\frac{x^2}{2} - \frac{1}{8}xy\Big|_0^1 = \frac{5}{8} - \frac{1}{8}y \text{ for } 0 < y < 2$$

which are both triangular probability density functions. Then

$$h_X = -\int_0^1 \left(\frac{7}{4} - \frac{3}{2}x\right)\ln\left(\frac{7}{4} - \frac{3}{2}x\right)dx$$

$$= -\frac{2}{3}\int_{\frac{1}{4}}^{\frac{7}{4}} w\ln(w)dw \text{ (by letting } w = \frac{7}{4} - \frac{3}{2}x)$$

$$= -\frac{2}{3}\left(\frac{w^2}{2}\ln(w) - \frac{w^2}{4}\right)\Big|_{\frac{1}{4}}^{\frac{7}{4}}$$

$$= -\frac{2}{3}\left(\frac{49}{32}\ln\left(\frac{7}{4}\right) - \frac{49}{64} - \frac{1}{32}\ln\left(\frac{1}{4}\right) + \frac{1}{64}\right) = -0.100 \text{ nats}$$

so, dividing by $\ln(2)$

$$h_X = -0.144 \text{ bits}.$$

Moreover, we have that

$$h_Y = -\int_0^2 \left(\frac{5}{8} - \frac{1}{8}y\right) \ln\left(\frac{5}{8} - \frac{1}{8}y\right) dy$$

$$= -8 \int_{\frac{3}{8}}^{\frac{5}{8}} w \ln(w) dw \text{ (by letting } w = \frac{5}{8} - \frac{1}{8}y)$$

$$= -8 \left(\frac{w^2}{2} \ln(w) - \frac{w^2}{4}\right) \Big|_{\frac{3}{8}}^{\frac{5}{8}}$$

$$= -8 \left(\frac{25}{128} \ln\left(\frac{5}{8}\right) - \frac{25}{256} - \frac{9}{128} \ln\left(\frac{3}{8}\right) + \frac{9}{256}\right) = 1.786 \text{ nats}$$

and so

$$h_Y = 2.577 \text{ bits}.$$

Furthermore, the joint entropy is given by

$$h_{XY} = -\int_0^2 \int_0^1 \left(1 - \frac{y}{8} - \frac{3}{4}x\right) \ln\left(1 - \frac{y}{8} - \frac{3}{4}x\right) dx dy$$

$$= -\int_0^2 \frac{4}{3} \int_{\frac{1}{4}-\frac{y}{8}}^{1-\frac{y}{8}} u \ln(u) du dy \text{ (where } u = \left(1 - \frac{y}{8}\right) - \frac{3}{4}x)$$

$$= -\int_0^2 \frac{4}{3} \left(\frac{u^2}{2} \ln(u) - \frac{u^2}{4}\right) \Big|_{\frac{1}{4}-\frac{y}{8}}^{1-\frac{y}{8}} dy$$

$$= -\frac{4}{3} \int_0^2 \left[\frac{1}{2}\left(1 - \frac{y}{8}\right)^2 \ln\left(1 - \frac{y}{8}\right) - \frac{1}{4}\left(1 - \frac{y}{8}\right)^2\right.$$

$$\left. - \frac{1}{2}\left(\frac{1}{4} - \frac{y}{8}\right)^2 \ln\left(\frac{1}{4} - \frac{y}{8}\right) + \frac{1}{4}\left(\frac{1}{4} - \frac{y}{8}\right)^2\right] dy$$

$$= -\frac{4}{3} \left[\frac{8}{2} \int_{\frac{3}{4}}^1 v^2 \ln(v) dv - \frac{8}{4} \int_{\frac{3}{4}}^1 v^2 dv - \frac{8}{2} \int_0^{\frac{1}{4}} w^2 \ln(w) dw + \frac{8}{4} \int_0^{\frac{1}{4}} w^2 dw\right]$$

where $v = 1 - \frac{y}{8}$ and $w = \frac{1}{4} - \frac{y}{8}$

$$= -\frac{4}{3} \left[4 \left(\frac{v^3}{3} \ln(v) - \frac{v^3}{9}\right) \Big|_{\frac{3}{4}}^1 - 2\frac{v^3}{3} \Big|_{\frac{3}{4}}^1 - 4 \left(\frac{w^3}{3} \ln(w) - \frac{w^3}{9}\right) \Big|_0^{\frac{1}{4}}\right.$$

$$\left. +2\frac{w^3}{3} \Big|_0^{\frac{1}{4}}\right] = 0.579 \text{ nats}$$

so $h_{XY} = 0.835$ bits.

Finally, the mutual information is given by

$$I_{XY} = h_X + h_Y - h_{XY} = 1.598 \text{ bits} \tag{2.21}$$

∎

In Chapter 4 we provide a detailed calculation of differential entropy for the probability density functions typically listed in a table of statistical distributions. Many of these results may also be found in the literature (see Lazo & Rathie [24]; also Cover & Thomas [8], pp. 486-487), but these references only provide computational hints. It is instructive in understanding the concept of differential entropy to see the complete proofs. A complete listing of the probability distributions considered in this work and their associated entropies can be found in Table 2.3.

A brief word on notation. Note that in Table 2.3 and in the associated derivations we drop the subscript X from the probability distribution functions for notational convenience. It will always be obvious from context that when we write $p(x)$ we are referring to the distribution function $p_X(x)$ associated with the random variable X. Similarly we write $p(x,y) \equiv p_{XY}(x,y)$ and $p(y|x) \equiv p_Y(y|x)$. Analogously for discrete random variables we use $f(x)$, $F(x)$ for the PMF and CDF, respectively.

2.3   Interpretation of Differential Entropy

The definition of differential entropy in Eqn. (2.9) appears to be simply a natural extension to continuous variables of the Shannon entropy for discrete variables in Eqn. (2.1). However, extreme care must be exercised in making this analogy. We have already pointed out some of the significant differences between the properties of these two entropies and, in this section, we will discuss these differences in greater depth.

A fundamental difference between the two definitions, and the cause of the necessity for different interpretations of the two quantities, arises from the fact that the $f_X(x_m)$ appearing in the (discrete) Shannon entropy (Eqn. 2.1) is a probability whereas the $p_X(x_m)$ appearing in the (continuous) differential entropy (Eqn. 2.9) is not. Instead, $p_X(x)$ is a probability density and acquires the meaning of probability only when it is integrated over a finite interval. The appearance of $p_X(x)$ all alone inside the log function is, therefore, open to wide interpretation. For discrete entropy, the interpretation as average uncertainty is meaningful since, as has been pointed out by Shannon [48], Jaynes [18], and others, the quantity $\log(1/f_x(x_m))$ is a logically consistent measure of uncertainty. For differential entropy, the corresponding claim cannot be made. In spite of these conceptual difficulties, differential entropy as a mathematical quantity finds wide utility in a number of important scientific disciplines, as will be summarized below. Importantly, the associated

TABLE 2.3
Probability density functions and the associated differential entropies.

Distribution	Probability density function	Differential entropy (in bits)
Beta	$p(x) = \begin{cases} \frac{\Gamma(\eta+\lambda)}{\Gamma(\eta)\Gamma(\lambda)} x^{\lambda-1}(1-x)^{\eta-1} & : 0 \leq x \leq 1 \\ 0 & : \text{otherwise} \end{cases}$ where $\lambda > 0, \eta > 0$	$\log_2\left(\frac{B(\eta,\lambda)e^{(\eta+\lambda-2)\Psi(\eta+\lambda)}}{e^{(\lambda-1)\Psi(\lambda)}e^{(\eta-1)\Psi(\eta)}}\right)$
Cauchy	$p(x) = \dfrac{1}{\pi b\left[1+\frac{(x-a)^2}{b^2}\right]}$ $-\infty < x < \infty$ where $b > 0$	$\log_2(4\pi b)$
Chi	$p(x) = \begin{cases} \frac{2(n/2)^{n/2}}{\sigma^n\Gamma(n/2)} x^{n-1} e^{-(n/(2\sigma^2))x^2} & : 0 \leq x < \infty \\ 0 & : \text{otherwise} \end{cases}$ where $\sigma > 0$ and n is a positive integer	$\log_2\left(\frac{\Gamma(n/2)\sigma}{\sqrt{2n}}e^{[n-(n-1)\Psi(n/2)]/2}\right)$
Chi-squared	$p(x) = \begin{cases} \frac{1}{2^{n/2}\Gamma(n/2)} x^{n/2-1} e^{-x/2} & : 0 < x < \infty \\ 0 & : \text{otherwise} \end{cases}$ where n is a positive integer	$\log_2\left(2\Gamma(n/2)e^{\frac{n}{2}-\frac{n-2}{2}\Psi(n/2)}\right)$
Dirac Delta Function	$p(x) = \delta(x)$	$-\infty$

-Continued-

TABLE 2.3
Probability density functions and the associated differential entropies. (continued)

Exponential	$p(x) = \begin{cases} \lambda e^{-\lambda x} : x \geq 0 \\ 0 : \text{otherwise} \end{cases}$ where $\lambda > 0$	$\log_2\left(\frac{e}{\lambda}\right)$
F-distribution	$p(x) = \begin{cases} \dfrac{v^{\frac{v}{2}} w^{\frac{w}{2}}}{B\left(\frac{v}{2},\frac{w}{2}\right)} \dfrac{x^{\frac{v}{2}-1}}{(w+vx)^{(v+w)/2}} : 0 \leq x < \infty \\ 0 : \text{otherwise} \end{cases}$ where v, w are positive integers	$\log_2\left(\dfrac{w}{v}B\left(\tfrac{v}{2},\tfrac{w}{2}\right)\dfrac{e^{(1-\frac{v}{2})\Psi(\frac{v}{2})}e^{(\frac{v+w}{2})\Psi(\frac{v+w}{2})}}{e^{(1+\frac{w}{2})\Psi(\frac{w}{2})}}\right)$
Gamma	$p(x) = \begin{cases} \dfrac{\lambda^\eta}{\Gamma(\eta)}x^{\eta-1}e^{-\lambda x} : 0 < x < \infty \\ 0 : \text{otherwise} \end{cases}$ where $\lambda > 0,\ \eta > 0$	$\log_2\left(\dfrac{\Gamma(\eta)}{\lambda}e^{\eta+(1-\eta)\Psi(\eta)}\right)$
Generalized Beta	$p(y) = \begin{cases} \dfrac{1}{b-a}\dfrac{\Gamma(\eta+\lambda)}{\Gamma(\eta)\Gamma(\lambda)}\left(\dfrac{y-a}{b-a}\right)^{\lambda-1}\left(\dfrac{b-y}{b-a}\right)^{\eta-1} : a \leq y \leq b \\ 0 : \text{otherwise} \end{cases}$ where $\lambda > 0,\ \eta > 0,\ a \geq 0$	$\log_2\left(\dfrac{(b-a)B(\eta,\lambda)e^{(\eta+\lambda-2)\Psi(\eta+\lambda)}}{e^{(\lambda-1)\Psi(\lambda)}e^{(\eta-1)\Psi(\eta)}}\right)$
Generalized Normal	$p(x) = \dfrac{\beta}{2\alpha\Gamma\left(\frac{1}{\beta}\right)}e^{-(|x-\mu|/\alpha)^\beta}$ $\qquad -\infty < x < \infty$ where $\alpha > 0,\ \beta > 0$	$\log_2\left(\dfrac{2\alpha\Gamma\left(\frac{1}{\beta}\right)}{\beta}e^{\frac{1}{\beta}}\right)$
Kumaraswamy	$p(x) = \begin{cases} abx^{a-1}(1-x^a)^{b-1} : 0 \leq x \leq 1 \\ 0 : \text{otherwise} \end{cases}$ where $a > 0,\ b > 0$	$\log_2\left(\dfrac{1}{ab}e^{\frac{b-1}{b}+\frac{a-1}{a}\left(\gamma+\Psi(b)+\frac{1}{b}\right)}\right)$

-Continued-

TABLE 2.3

Probability density functions and the associated differential entropies. (continued)

Laplace	$p(x) = \frac{1}{2}\lambda e^{-\lambda\|x\|}$ $-\infty < x < \infty$ where $\lambda > 0$	$\log_2\left(\frac{2e}{\lambda}\right)$
Log Normal	$p(x) = \begin{cases} \frac{1}{\sqrt{2\pi}\sigma x}e^{-\left(\ln(x)-\ln(m)\right)^2/2\sigma^2} : 0 < x < \infty \\ 0 : \text{otherwise} \end{cases}$ where $m>0,\ \sigma>0$	$\frac{1}{2}\log_2\left(2\pi e\sigma^2 m^2\right)$
Logistic	$p(x) = \frac{e^{-(x-\mu)/s}}{s\left(1+e^{-(x-\mu)/s}\right)^2}$ $-\infty < x < \infty$ where $s>0$	$\log_2(se^2)$
Log-logistic	$p(x) = \begin{cases} \frac{\frac{\beta}{\alpha}\left(\frac{x}{\alpha}\right)^{\beta-1}}{\left[1+\left(\frac{x}{\alpha}\right)^\beta\right]^2} : 0 \le x < \infty \\ 0 : \text{otherwise} \end{cases}$ where $\alpha>0,\ \beta>0$	$\log_2\left(\frac{\alpha}{\beta}e^2\right)$
Maxwell	$p(x) = \begin{cases} \frac{4}{\sqrt{\pi}}\frac{x^2 e^{-x^2/(\alpha^2)}}{\alpha^3} : 0 \le x < \infty \\ 0 : \text{otherwise} \end{cases}$ where $\alpha>0$	$\log_2\left(\sqrt{\pi}\alpha e^{\gamma-1/2}\right)$
Mixed Gaussian	$p(x) = \frac{1}{2\sigma\sqrt{2\pi}}\left[e^{-(x-\mu)^2/2\sigma^2} + e^{-(x+\mu)^2/2\sigma^2}\right]$ $-\infty < x < \infty$ where $\mu \ge 0, \sigma > 0$	$\frac{1}{2}\log_2(2\pi e\sigma^2) + L(\mu/\sigma)$ where $L(\cdot)$ is a monotonically increasing function from 0 to 1(see 7.4)

-Continued-

TABLE 2.3
Probability density functions and the associated differential entropies. (continued)

Nakagami $\quad p(x) = \begin{cases} \frac{2}{\Gamma(m)} \left(\frac{m}{\Omega}\right)^m x^{2m-1} e^{-mx^2/\Omega} & : x > 0 \\ 0 & : \text{otherwise} \end{cases}$ $\quad \log_2 \left(\frac{\Gamma(m)}{2} \sqrt{\frac{\Omega}{m}} e^{[2m-(2m-1)\Psi(m)]/2}\right)$

where $\Omega > 0$ and $m \geq \frac{1}{2}$

Normal $\quad p(x) = \frac{1}{\sigma\sqrt{2\pi}} e^{-(x-\mu)^2/2\sigma^2} \qquad -\infty < x < \infty$ $\qquad \frac{1}{2}\log_2(2\pi e \sigma^2)$

where $\sigma > 0$

Pareto $\quad p(x) = \begin{cases} c x_o^c x^{-(c+1)} & : x_o \leq x < \infty \\ 0 & : \text{otherwise} \end{cases}$ $\qquad \log_2 \left(\frac{x_o}{c} e^{1+\frac{1}{c}}\right)$

where $c > 0$, $x_o \geq 1$

Rayleigh $\quad p(x) = \begin{cases} \frac{x}{\sigma^2} e^{-x^2/2\sigma^2} & : 0 \leq x < \infty \\ 0 & : \text{otherwise} \end{cases}$ $\qquad \log_2 \left(\frac{\sigma}{\sqrt{2}} e^{1+\frac{\gamma}{2}}\right)$

where $\sigma > 0$

Rice $\quad p(x) = \begin{cases} \frac{x}{\sigma^2} e^{-(x^2+a^2)/2\sigma^2} I_0\left(\frac{ax}{\sigma^2}\right) & : x \geq 0 \\ 0 & : \text{otherwise} \end{cases}$ $\qquad \log_2 \left(\frac{\sigma}{\sqrt{2}} e^{1+\frac{\gamma}{2}}\right) + \frac{a^2}{\sigma^2 \ln(2)}$

(see derivation) where $a \geq 0$, $\sigma > 0$ $\qquad$ + integral functions of a and σ

where I_0 is the modified

Bessel function of the first kind, order 0

Simpson $\quad p(x) = \begin{cases} \frac{a-|x|}{a^2} & : -a \leq x \leq a \\ 0 & : \text{otherwise} \end{cases}$ $\qquad \log_2(a\sqrt{e})$

-Continued-

TABLE 2.3
Probability density functions and the associated differential entropies. (continued)

Sine Wave $p(x) = \begin{cases} \dfrac{1}{\pi\sqrt{A^2-x^2}} & : -A < x < A \\ 0 & : \text{otherwise} \end{cases}$ $\log_2\left(\dfrac{\pi A}{2}\right)$

where $A > 0$

Students-t $p(x) = \dfrac{1}{\sqrt{n}B\left(\frac{1}{2},\frac{n}{2}\right)}\left(1+\dfrac{x^2}{n}\right)^{-(n+1)/2}$ $-\infty < x < \infty$ $\log_2\left(\sqrt{n}B\left(\frac{1}{2},\frac{n}{2}\right)e^{\frac{(n+1)}{2}\left[\Psi\left(\frac{n+1}{2}\right)-\Psi\left(\frac{n}{2}\right)\right]}\right)$

where n = positive integer

Truncated $p(x) = \begin{cases} \dfrac{1}{Z\sigma\sqrt{2\pi}}e^{-(x-\mu)^2/2\sigma^2} & : a \le x \le b \\ 0 & : \text{otherwise} \end{cases}$ $\log_2\left(\sqrt{2\pi e}\sigma Z e^{\frac{1}{2Z}}\right)$
Normal
$$\times\left[\left(\dfrac{a-\mu}{\sigma}\right)\phi\left(\dfrac{a-\mu}{\sigma}\right) - \left(\dfrac{b-\mu}{\sigma}\right)\phi\left(\dfrac{b-\mu}{\sigma}\right)\right]$$

where $Z = P\left(\dfrac{b-\mu}{\sigma}\right) - P\left(\dfrac{a-\mu}{\sigma}\right)$ $a < \mu < b,\ \sigma > 0$

$P \equiv$ Standard Normal CDF $\phi \equiv$ Standard Normal PDF

Uniform $p(x) = \begin{cases} \dfrac{1}{a} & : 0 < x \le a \\ 0 & : \text{otherwise} \end{cases}$ $\log_2(a)$

Weibull $p(x) = \begin{cases} \dfrac{\eta}{\sigma}\left(\dfrac{x}{\sigma}\right)^{\eta-1}e^{-\left(\frac{x}{\sigma}\right)^{\eta}} & : 0 \le x < \infty \\ 0 & : \text{otherwise} \end{cases}$ $\log_2\left(\dfrac{\sigma}{\eta}e^{1+\frac{\eta-1}{\eta}\gamma}\right)$

where $\eta > 0,\ \sigma > 0$

continuous forms of mutual information and transfer entropy retain their interpretation from the discrete case. These two forms comprise *differences* of individual (or conditional) entropies and here we are on very safe ground. In dealing with individual differential entropies, the absolute value is almost never of interest. Rather the quantities of interest are often either the differences between differential entropy for the same density but under different conditions, or a comparison of the differential entropy for different probability densities under common conditions. Some examples are given below.

Consider measurements on a system that outputs voltages over the continuous range between $0.00V$ and $1.00V$ and in which the probability density for this variable is uniform over the range of voltages. Using Table 2.3 we find that the differential entropy for a uniform distribution on the interval $(0, a]$ is $h_X = \log_2(a)$. Hence, if the results of the voltage measurements are specified in volts we have $h_X = \log_2(1) = 0$ *bits*. However, if the results of the same voltage measurement are specified instead in milliVolts, then $h_X = \log_2(1000) \approx 9.97$ *bits*. Suppose now that a second experiment was performed in which the voltage range was doubled to $a = 2.00V$ and in which it was known that the pdf was still uniform. The entropy values now become $h_X = \log_2(2) = 1$ *bit* and $h_X = \log_2(2000) \approx 10.97$ *bits*. In each case, the *change* in entropy was 1 bit, regardless of the choice of units. Changes in entropy thus do not depend on the choice of units.

Now suppose that, in the second experiment, the voltage range remained at $1V$ but the density changed to a generalized beta distribution with parameters $\lambda = \eta = 2$, $a = 0$, $b = 1$. Using Table 2.3, we find

$$h_X = \log_2(b - a) + \log_2\left(\frac{B(2, 2)\exp(2\Psi(4))}{\exp(2\Psi(2))}\right)$$
$$= \log_2(b - a) - 0.18 \ bits \tag{2.22}$$

Notice that the leading term on the right side is just the entropy for the uniform distribution over the same range. The second term, therefore, represents the change, in this case the reduction, in the entropy due to the change in *pdf*. Note also that the second term depends only on parameters associated with the shape of the distribution and not on the range (support) of the distribution, as must be the case to guarantee that the change is invariant to scaling of the range.

The above examples bring out a practical issue with differential entropy that is rarely discussed explicitly in the literature. Many of the continuous *pdfs* summarized in Table 2.3 are used everyday to model or describe actual physical experiments, in which case the *pdf* can indeed have units. Taking the simplest example, if a voltage measurement is modeled by a uniform distribution on the interval $[a, b]$ volts, then the *pdf* is $p_V(v) = (1/(b - a))V^{-1}$. Note that the normalization condition is still automatically satisfied

$$\int_a^b \left(\frac{1}{(b - a)}\right) dv = 1. \tag{2.23}$$

However, a quantity with units cannot form the argument of the log function. Hence, any continuous *pdf* used in the calculation of differential entropy must first be cast in dimensionless form. In our example of the uniform voltage *pdf*, this can be done by replacing the voltage variable v by the new variable $x = v/v_{ref}$ where v_{ref} is a suitably chosen reference voltage. The uniform pdf defined on this new variable is clearly dimensionless. A clear choice for the reference value is $v_{ref} = 1V$ in which case x is simply the numerical value of the voltage v in Volts. Obviously, when comparing the differential entropy for two or more densities, the same reference value must be used.

A clue that only dimensionless variables are acceptable is found in the form of the densities for the beta and the Kumaraswamy distributions. An inspection of these densities shows that the random variable x must be dimensionless. Any experimental outcome modeled by one of these densities must first be put in dimensionless form, independent of any consideration of the entropy.

We make one final point. The change of variables $v \rightarrow v/v_{ref}$ is made prior to writing the *pdf*. It is not made in the integral expression for differential entropy. To make the change in the integral itself would require first inserting a dimensioned *pdf* into the log function, and no variable change after that can overcome the initial difficulty.

2.4   Historical and Scientific Perspective

The notion of entropy first appeared in the mid-19th century by Rudolph Clausius in the context of thermodynamics. Researchers in this time period were forced to understand and predict macroscopic observables using only macroscopic concepts, since the atomic (microscopic) nature of matter was not yet understood. In seeking a more fundamental understanding of the physics underlying heat engines, Clausius showed that there exists an exact differential function $dS = dQ/T$ of any system such that, whenever the system undergoes a closed loop, reversible process, the entropy remains unchanged $dS = 0$. Here dQ is the heat transferred into the system at constant temperature T. Clausius further showed that in any equilibrium state, the observable thermodynamic state parameters (such as pressure, volume, mole number) attain values for which the entropy is maximum subject to any physical constraints. (See, e.g., [7, 52]). Later in the 19th and early 20th centuries, as the atomic hypothesis was gaining wide acceptance, James Clerk Maxwell, Ludwig Boltzmann and Joshua Willard Gibbs were formulating the principles of statistical mechanics. Now scientists had the advantage of employing microscopic concepts to predict macroscopic parameters. Just as in classical thermodynamics, the goal was to calculate the values of observable physical quantities but this time with the understanding that the thermodynamic system comprised a collection of a large number of particles — atoms and molecules. For a system containing N particles, all possible states of the system were assumed to exist as points in a real, continuous 6N-dimensional space — the so-called phase space of the system. In order to calculate observable quantities, what was needed was the probability density $f(\vec{r}, \vec{p}, t)$ in phase space defined such that $f(\vec{r}, \vec{p}, t)d^3\vec{r}d^3\vec{p}$ is the fraction of particles that, at time t, have positions in the volume element $d^3\vec{r}$ about $\vec{r}$ and momentum in the volume element $d^3\vec{p}$ about $\vec{p}$. Then the correct distribution is the one that maximizes the functional

$$H(t) = -k_B \int d^3\vec{p} f(\vec{p}, t) \log(f(\vec{p}, t)) \qquad (2.24)$$

subject to any known constraints, where k_B is the Boltzmann constant. We recognize this expression today as the differential entropy of the phase space probability density. (Notice that $\vec{r}$ no longer appears in the distribution. In the absence of external forces it is reasonable to assume that the distribution is independent of position.) As the 20th century progressed, it became clear that the "classical" atomic view of the world was itself insufficient and that the more accurate model of the physical world was the quantum theory. In employing this new model to the calculation of thermodynamic quantities, the overall approach was similar to the phase space model described above. Now, however, phase space is no longer continuous but is quantized, the random variables are now discrete, and the function to be maximized subject to constraints is the Shannon entropy multiplied by the Boltzmann constant k_B.

$$S = -k_B \sum_i f_i \log(f_i). \qquad (2.25)$$

In the discussion above we have described what is essentially the principle of maximum entropy: the correct distribution for describing a physical system is found by maximizing the entropy (either discrete or continuous) subject to any constraints imposed by the physics of the situation [17]. To perform the calculation, the method of Lagrange multipliers is applied in a fairly straightforward manner. Once the probability distribution $f_X(x)$ is obtained, the average of any observable that is a function g of the random variable X is just the expectation of g with respect to the distribution, $\langle g \rangle = E[g] \equiv \sum_i g(x_i) f_X(x_i)$.

This completes our very brief description of the central place held by the concept of entropy in modern statistical mechanics. The next significant development in entropy came in 1948 from an unexpected area: communication theory. In that year, Claude Shannon developed a complete theoretical framework for quantifying the performance of communication links.

Any communication link consists of a source, or transmitter, a physical link, and a receiver. Such a link will be used to transmit messages of interest. In any practical system, the messages are first encoded onto a finite set, or alphabet, of symbols. Let the alphabet be denoted by A, assume it consists of N different symbols $A = \{a_1, \cdots, a_N\}$, and assume the corresponding probability distribution is $f_A(a_i)$. Shannon showed the importance of the entropy function

$$H_A = -\sum_i f_A(a_i) \log_2 (f_A(a_i)) \qquad (2.26)$$

in predicting link performance. Here Shannon interpreted H_A as the average uncertainty, in bits per symbol, of the alphabet A and referred to it as the entropy of the source. Equivalently, it is the average number of bits per symbol needed to optimally encode the alphabet A prior to transmission through the link. He showed that the theoretical capacity C of the link, in bits, is given by $C = H_A$. If the source emits ρ symbols per second, then we can say that the entropy of the source and, hence, the link capacity, in bits per second, is $C = \rho H_A$. Shannon showed that effectively error-free transmission is guaranteed in this discrete channel provided the system transmits symbols at rate R no faster than C, $R \leq C$.

The reader may rightfully raise a serious question at this point. Earlier in the chapter we argued for interpreting entropy as a measure of uncertainty in our state of knowledge. In the case of a noise-free communications system, how can there possibly be any uncertainty associated with the source? Presumably the sender knows exactly what he or she wishes to transmit. So where is the uncertainty? As correctly pointed out by Jaynes [18], the uncertainty resides with the engineer designing the system! It is the system designer who knows only the probability distribution associated with the source alphabet and Shannon has shown that it is this knowledge, and not the knowledge of any particular message, that is sufficient to quantify the performance of the system.

Suppose now that the system is contaminated with noise so that when the symbol a_i is sent, the same symbol is not necessarily received. To keep track of things, let $f_B(b_k)$ and H_B be the probability distribution and entropy, respectively, associated with the receiver. Let $H_{B|A}$ be the conditional entropy associated with the conditional probability $f_B(b_k|a_i)$, that is, with the probability that b_k was received given that a_i was sent. Compared to the noise-free case, the theoretical channel capacity must be smaller. In fact it is now $H_A - H_{B|A}$. Effectively error-free transmission is still possible in this case but the symbol transmission rate R must be reduced to $R \leq (H_A - H_{B|A})$. Note that the right side of this equation is just the mutual information between the source and the receiver.

The concept of differential entropy has found applicability in fields well beyond statistical mechanics and communication theory. These include finance and econometrics, spectroscopy, image processing, population research and ecology, and biostatistics. Following the work of Jaynes [18] cited earlier, it is now recognized that the great utility of entropy arises, not in its own right, but when combined with Bayes' theorem and the principle of entropy maximization. Here, Bayes' theorem provides the recipe for calculating a desired (a posteriori) distribution from a known or given (a priori) distribution, and entropy maximization provides the principle for obtaining the a priori distribution provided the constraints of the problem are known accurately.

3

Entropy for Discrete Probability Distributions

3.1 Poisson Distribution

The Poisson distribution represents the probability of exactly x events occurring in a specified interval, based on a rate of occurrence λ. Its probability mass function is given by

$$f(x) = \frac{\lambda^x e^{-\lambda}}{x!}, \qquad x = 0, 1, 2, 3, \cdots ; \ \lambda > 0 \tag{3.1}$$

The mean of the Poisson distribution is λ, and the variance is also given by λ. The discrete entropy for various rates λ is given in Figure 3.1. As expected, the value of the entropy increases as the variance λ increases.

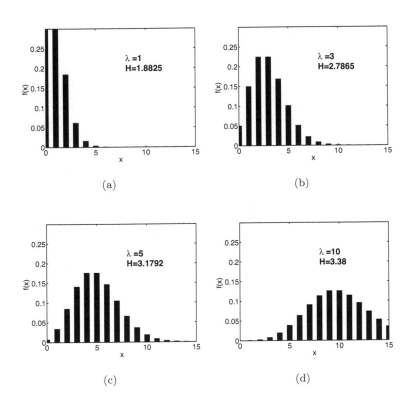

FIGURE 3.1
Poisson distribution and the associated discrete entropy for various values of the rate parameter λ.

3.2 Binomial Distribution

The binomial distribution represents the probability of x successes in n independent (Bernoulli) trials, with sampling from an infinite population. The probability mass function is given by

$$f(x) = \binom{n}{x} p^x (1-p)^{n-x} \qquad x = 0, 1, 2, \cdots, n; \ 0 \le p \le 1 \qquad (3.2)$$

The mean of this distribution is np and the variance is $np(1-p)$. The entropy, for various values of p and n is shown in Figure 3.2.

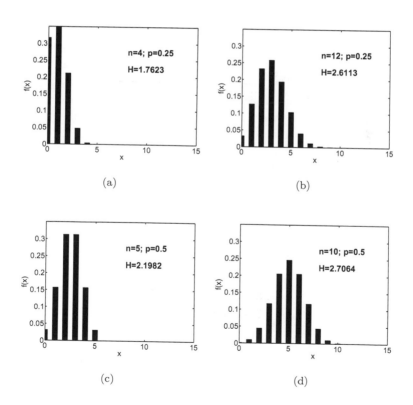

FIGURE 3.2

Binomial distribution and the associated discrete entropy for various values of the parameters n and p.

3.3   Hypergeometric Distribution

The hypergeometric distribution represents the probability of x successes in n trials, when n is not small relative to the population size N of which K units in the population are known to be successful. The probability mass function is given by

$$f(x) = \frac{\binom{K}{x}\binom{N-K}{n-x}}{\binom{N}{n}} \qquad x = 0, 1, 2, \cdots, n$$

$$\text{where } x \leq K, \ n - x \leq N - K \qquad (3.3)$$

The mean of this distribution is $\frac{nK}{N}$ and the variance is given by

$$\sigma^2 = \frac{nK(N-K)(N-n)}{N^2(N-1)}$$

The PMF and the associated entropy is shown in Figure 3.3 for various values of N, K, n.

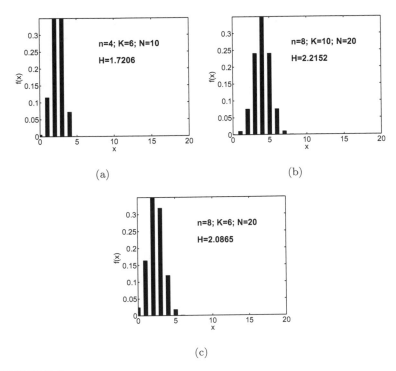

(a)   (b)

(c)

FIGURE 3.3
Hypergeometric distribution and the associated discrete entropy for various values of parameters N, K, n.

3.4 Geometric Distribution

The geometric distribution represents the probability that exactly x independent (Bernoulli) trials are required until the first success is achieved. The
probability mass function is given by

$$f(x) = (1-p)^{x-1}p \qquad x = 1, 2, \cdots ; \ 0 \le p \le 1 \qquad (3.4)$$

The mean of this distribution is $\frac{1}{p}$ and the variance is given by $\sigma^2 = \frac{1-p}{p^2}$.
Entropy is shown in Figure 3.4 for various values of p.

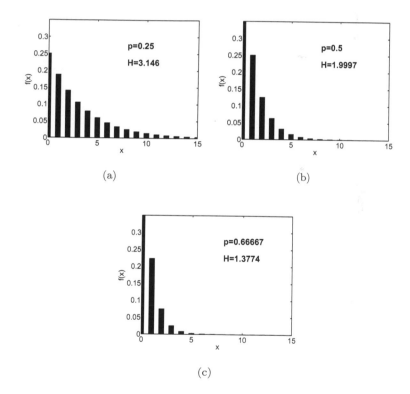

(a)   (b)

(c)

FIGURE 3.4
Geometric distribution and the associated discrete entropy for various values
of parameters p.

3.5   Negative Binomial Distribution

The negative binomial distribution represents the probability that if independent (Bernoulli) trials are conducted until exactly s successes are obtained, then there are x failures (so there are $s + x$ total trials). The probability mass function is given by:

$$f(x) = \binom{x + s - 1}{x} p^s (1 - p)^x \qquad x = 0, 1, 2, \cdots ; \; 0 \leq p \leq 1 \qquad (3.5)$$

where s is a positive integer. The mean of this distribution is $\mu = \frac{s(1-p)}{p}$ and the variance is $\sigma^2 = \frac{s(1-p)}{p^2}$. Entropy is shown in Figure 3.5 for various values of p and s.

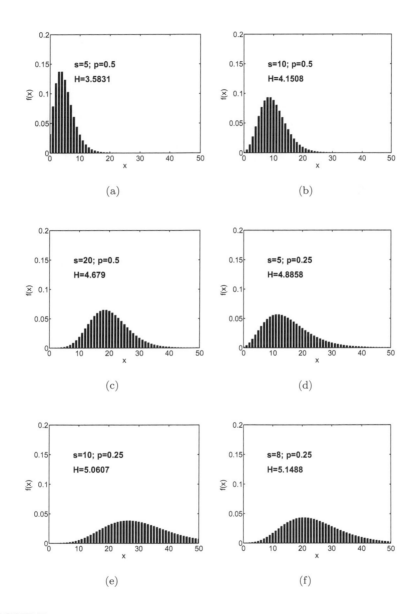

FIGURE 3.5
Negative binomial distribution and the associated discrete entropy for various values of parameters s, p.

3.6   Discrete Uniform Distribution

This distribution is the discrete version of the continuous uniform distribution. The probability function is simply

$$f(x) = \frac{1}{b} \qquad x = a, a+1, \cdots, a + (b-1) \tag{3.6}$$

where b is a positive integer. The mean is $\mu = a + \frac{b-1}{2}$ and the variance is $\sigma^2 = \frac{(b^2-1)}{12}$. The entropy is $H = \log_2(b)$ bits, independently of the value of a.

3.7 Logarithmic Distribution

The logarithmic distribution is derived from the MacLaurin series expansion

$$-\ln(1-p) = p + \frac{p^2}{2} + \frac{p^3}{3} + \cdots$$

which, when rearranged, gives

$$\sum_{n=1}^{\infty} \frac{-1}{\ln(1-p)} \frac{p^n}{n} = 1$$

and therefore

$$f(x) = \frac{-1}{\ln(1-p)} \frac{p^x}{x} \qquad x = 1, 2, 3, \cdots; \; 0 < p < 1 \qquad (3.7)$$

is a probability mass function. The distribution has as its first two moments

$$\mu = \frac{-1}{\ln(1-p)} \frac{p}{1-p}$$

$$\sigma^2 = \frac{-p\,(p + \ln(1-p))}{(1-p)^2\,[\ln(1-p)]^2}.$$

The entropy is shown in Figure 3.6 for various values of p.

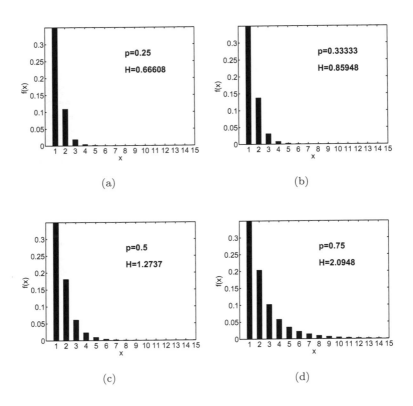

FIGURE 3.6
Logarithmic distribution and the associated discrete entropy for various values of the parameter p.

3.8 Skellam Distribution

The Skellam distribution arises as the distribution of the difference of two independent Poisson random variables with different expected values. It is useful in describing point spreads in sports where all individual scores are equal, such as baseball and hockey (but not, for example, football or basketball). The probability mass function is given by

$$f(x) = e^{-(\mu_1 + \mu_2)} \left(\frac{\mu_1}{\mu_2} \right)^{x/2} I_{|x|} \left(2\sqrt{\mu_1 \mu_2} \right) \ x = \cdots, -2, -1, 0, 1, 2, \cdots$$

$$\mu_1 \geq 0, \ \mu_2 \geq 0 \tag{3.8}$$

where I_k is the modified Bessel function of the first kind. The mean for this distribution is $\mu_1 - \mu_2$ and the variance is $\mu_1 + \mu_2$, since it is derived for the difference of two independent Poisson variables. The entropy is shown in Figure 3.7 for various values of μ_1, μ_2.

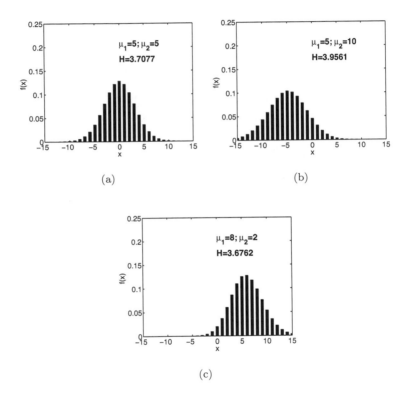

FIGURE 3.7

Skellam distribution and the associated discrete entropy for various values of parameters μ_1, μ_2.

3.9   Yule-Simon Distribution

The Yule-Simon distribution, named after Udny Yule and Herbert A. Simon, arose in the study of certain biological stochastic processes. The probability mass function is given by

$$f(x) = \rho B\left(x, \rho + 1\right) \qquad x = 1, 2, 3, \cdots ; \ \rho > 0 \qquad (3.9)$$

where $B(\cdot, \cdot)$ is the Beta function. This distribution has

$$\mu = \frac{\rho}{\rho - 1} \text{ for } \rho > 1$$

$$\sigma^2 = \frac{\rho^2}{(\rho - 1)^2(\rho - 2)} \text{ for } \rho > 2$$

Entropy is shown in Figure 3.8 for various values of ρ.

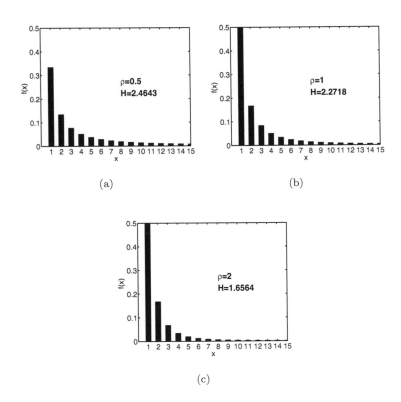

FIGURE 3.8
Yule-Simon distribution and the associated discrete entropy for various values
of the parameter ρ.

3.10   Zeta Distribution

The zeta distribution obtains its name from its use of the Riemann zeta function

$$\zeta(s) = \sum_{n=1}^{\infty} \frac{1}{n^s} \tag{3.10}$$

which is finite when $s > 1$. The probability mass function for the zeta distribution is given by

$$f(x) = \frac{1}{x^s \zeta(s)} \qquad x = 1, 2, 3, \cdots ; \ s > 1. \tag{3.11}$$

This distribution has

$$\mu = \frac{\zeta(s-1)}{\zeta(s)} \text{ for } s > 2$$

$$\sigma^2 = \frac{\zeta(s)\zeta(s-2) - [\zeta(s-1)]^2}{[\zeta(s)]^2} \text{ for } s > 3$$

$$H = \sum_{x=1}^{\infty} \frac{1}{x^s \zeta(s)} \log_2 \left(x^2 \zeta(s) \right)$$

This entropy is shown in Figure 3.9 for various values of s.

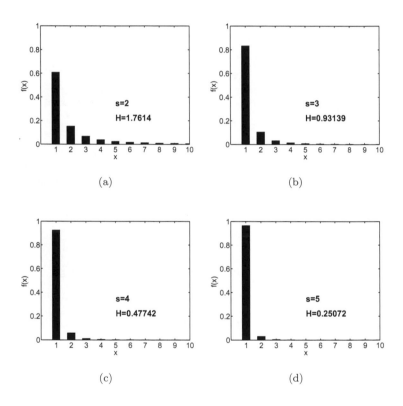

FIGURE 3.9
Zeta distribution and the associated discrete entropy for various values of the parameter s.

4

Differential Entropies for Probability Distributions

4.1 Beta Distribution

The Beta distribution is a two-parameter distribution which has a finite support set $[0, 1]$. Because it allows for a wide variety of shapes, it is the primary distribution used to model data with a defined maximum and minimum value. When $\eta > 1$ and $\lambda > 1$, the distribution is unimodal; when $\eta < 1$ and $\lambda < 1$, the distribution is U-shaped; when $\eta < 1$ and $\lambda \geq 1$ or $\lambda < 1$ and $\eta \geq 1$, the distribution is J-shaped or reverse J-shaped; when $\eta = \lambda$, the distribution is symmetric. When $\eta = \lambda = 1$, the distribution reduces to the uniform distribution. As can be seen from the graph, the differential entropy for the Beta distribution is always less than or equal to 0, reaching its maximum value of 0 when it reduces to the uniform distribution on $[0, 1]$. This is to be expected since the entropy is maximized when all events are equally probable.

The Beta distribution is closely related to the Kumaraswamy distribution. If X has a uniform distribution on $[0, 1]$, then X^2 has a Beta distribution. If X and Y are independent Gamma random variables, then $\frac{X}{X+Y}$ has a Beta distribution.

The Beta distribution is relevant to the following general problem. Suppose n independent random observations have been taken of some variable with unknown probability density function. Rank these observations in order from smallest to largest. It turns out that the proportion of the population which is between the kth smallest and m^{th} largest of the sample observations obeys a Beta distribution with $\lambda = n - k - m + 1$ and $\eta = k + m$. This remains valid regardless of the underlying density function. For example, suppose a lot of 60 light bulbs are tested until the time t, when the first failure is recorded. From this one can infer the probability that at least 95% of the light bulbs produced from this process will last beyond t, which is the reliability measure to be used to advertise the light bulbs. In this example, $k = 1$, $m = 0$, $\lambda = 60$, $\eta = 1$, and the probability is calculated to be $1 - F_\beta(0.95; 60, 1) = 0.954$ where $F_\beta(\cdot)$ is the incomplete Beta function.

The Beta distribution is also frequently used as a prior probability distribution in Bayesian estimation problems. Its usefulness in this regard stems from the fact that it is conjugate to (among others) the binomial probability mass function, $f_\Theta(\theta; k, N) = \theta^k(1 - \theta)^{N-k}$. The binomial distribution is a discrete distribution used frequently to predict the probability of having observed k successes in N trials given that the probability of success in any 1 trial is θ. It is often the case that we are given a dataset N, k with the goal of estimating θ. Taking a Bayesian approach, the estimated posterior probability distribution for θ is given by

$$p(\theta) = \frac{f_\Theta(\theta; k, N)p_B(\theta; \lambda, \eta)}{\int f_\Theta(\theta; k, N)p_B(\theta; \lambda, \eta)d\theta}$$
$$= p_B(\theta; \lambda + k, \eta + N - k) \tag{4.1}$$

where the notation $p_B(\theta; \lambda, \eta)$ denotes the Beta probability distribution for the random variable θ governed by parameters λ, η. That is to say, if one multiplies a binomial distribution with parameter θ by a beta distribution in θ and scales the result such that it integrates to unity (i.e., is a proper PDF), the result is also a Beta distribution with parameters modified by the observed data. Although numerical approaches to solving (4.1) exist, the closed-form expression offered by a Beta prior eliminates costly computations. Moreover, the flexibility offered by the Beta distribution makes it an attractive prior. As we have already pointed out, setting $\lambda = \eta = 1$ results in the uniform or uninformative prior. More examples are provided in the section on the Generalized Beta distribution.

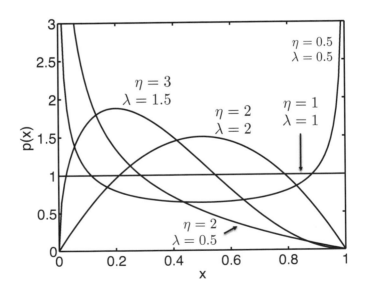

FIGURE 4.1
Probability density functions for the beta distribution.

Probability density: $p(x) = \begin{cases} \frac{\Gamma(\eta+\lambda)}{\Gamma(\eta)\Gamma(\lambda)} x^{\lambda-1}(1-x)^{\eta-1} & : \quad 0 \leq x \leq 1 \\ 0 & : \quad \text{otherwise} \end{cases}$

Range: $0 \leq x < 1$

Parameters: $\lambda > 0 \quad \eta > 0$

Mean: $\frac{\lambda}{\eta+\lambda}$

Variance: $\frac{\eta\lambda}{(\eta+\lambda)^2(\eta+\lambda+1)}$

r^{th} **moment about the origin:** $\prod_{i=0}^{r-1} \frac{\lambda+i}{\lambda+\eta+i}$

Mode: $\frac{\lambda-1}{\eta+\lambda+2} \quad$ if $\quad \lambda > 1 \quad$ and $\quad \eta > 1$

Characteristic function: $_1F_1(\lambda; \lambda + \eta; it)$

where $_1F_1(\cdot; \cdot; \cdot)$ is the Generalized Hypergeometric series.

Entropy: $h_X = \log_2\left(\frac{B(\eta,\lambda)e^{(\eta+\lambda-2)\Psi(\eta+\lambda)}}{e^{(\lambda-1)\Psi(\lambda)}e^{(\eta-1)\Psi(\eta)}}\right)$

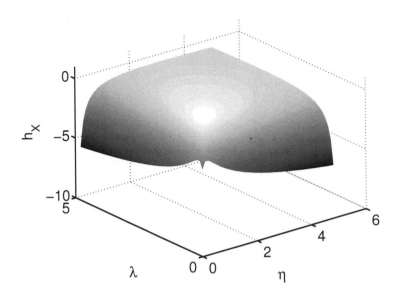

FIGURE 4.2
Differential entropy for the beta distribution.

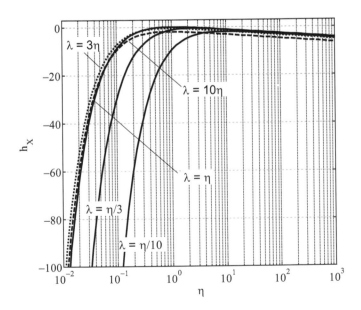

FIGURE 4.3
Differential entropy for the beta distribution for various $\frac{\lambda}{\eta}$ ratios.

Derivation of the differential entropy for the beta distribution

This case makes use of the following integral formula which is taken from Gradshteyn and Ryzhik [14], formula 4.253(1), pp. 538

$$\int_0^1 x^{\mu-1}(1-x^r)^{\nu-1}\ln(x)dx = \frac{1}{r^2}B\left(\frac{\mu}{r},\nu\right)\left[\Psi\left(\frac{\mu}{r}\right) - \Psi\left(\frac{\mu}{r}+\nu\right)\right]$$

provided that $Re(\mu) > 0$, $Re(\nu) > 0$, $r > 0$. Here, $B(\cdot,\cdot)$ is the Beta function and $\Psi(\cdot)$ is the Psi or Digamma function. Replacing $\frac{\Gamma(\eta)\Gamma(\lambda)}{\Gamma(\eta+\lambda)}$ by $B(\eta,\lambda)$ in the formula for the Beta distribution leads to

$$h_X = -\int_0^1 \frac{1}{B(\eta,\lambda)}x^{\lambda-1}(1-x)^{\eta-1}\ln\left[\frac{1}{B(\eta,\lambda)}x^{\lambda-1}(1-x)^{\eta-1}\right]dx$$

$$= -\int_0^1 \frac{1}{B(\eta,\lambda)}x^{\lambda-1}(1-x)^{\eta-1}[-\ln(B(\eta,\lambda)) + (\lambda-1)\ln(x)$$

$$+ (\eta-1)\ln(1-x)]dx$$

$$= \ln(B(\eta,\lambda))\int_0^1 \frac{1}{B(\eta,\lambda)}x^{\lambda-1}(1-x)^{\eta-1}dx$$

$$- \frac{1}{B(\eta,\lambda)}\left[(\lambda-1)\int_0^1 x^{\lambda-1}(1-x)^{\eta-1}\ln(x)dx\right.$$

$$\left. + (\eta-1)\int_0^1 x^{\lambda-1}(1-x)^{\eta-1}\ln(1-x)dx\right]$$

set $y = (1-x)$ in the third integral

$$= \ln(B(\eta,\lambda)) - \frac{1}{B(\eta,\lambda)}\left[(\lambda-1)\int_0^1 x^{\lambda-1}(1-x)^{\eta-1}\ln(x)dx\right.$$

$$\left. + (\eta-1)\int_0^1 y^{\eta-1}(1-y)^{\lambda-1}\ln(y)dy\right]$$

$$= \ln(B(\eta,\lambda)) - \frac{1}{B(\eta,\lambda)}[(\lambda-1)B(\lambda,\eta)(\Psi(\lambda) - \Psi(\lambda+\eta))$$

$$+ (\eta-1)B(\eta,\lambda)(\Psi(\eta) - \Psi(\eta+\lambda))]$$

By twice using the above integral with r=1

$$= \ln(B(\eta, \lambda)) - [(\lambda - 1)(\Psi(\lambda) - \Psi(\lambda + \eta)) + (\eta - 1)(\Psi(\eta) - \Psi(\eta + \lambda))]$$
$$= \ln(B(\eta, \lambda)) - (\lambda - 1)\Psi(\lambda) - (\eta - 1)\Psi(\eta) + (\eta + \lambda - 2)\Psi(\eta + \lambda)$$
$$= \ln\left(\frac{B(\eta, \lambda)e^{(\eta + \lambda - 2)\Psi(\eta + \lambda)}}{e^{(\lambda - 1)\Psi(\lambda)}e^{(\eta - 1)\Psi(\eta)}}\right)$$

so

$$h_X = \log_2\left(\frac{B(\eta, \lambda)e^{(\eta + \lambda - 2)\Psi(\eta + \lambda)}}{e^{(\lambda - 1)\Psi(\lambda)}e^{(\eta - 1)\Psi(\eta)}}\right) \qquad \text{bits}$$

4.2 Cauchy Distribution

The Cauchy distribution is named after Augustin Cauchy, but it is also sometimes referred to as the "Lorentz distribution." It occurs in a natural way: If X and Y are independent normally distributed random variables with zero means, then X/Y is Cauchy distributed. It also has physical applications; it arises in spectroscopy and in physics as the solution to the forced resonance differential equation. However, it is an unusual distribution for it has no mean; the mean and all higher order moments are undefined and there is no moment generating function. The mode, or median, is used as the measure of centrality, and the Cauchy distribution is symmetric about that mode. It is quite amazing that for such a "pathological" distribution the entropy has such a simple form.

The Cauchy distribution has some nice properties as well. If X is a Cauchy random variable, then so is kX and also $1/X$ if $a = 0$. If X and Y are independent Cauchy variables, then $X + Y$ is a Cauchy variable. In fact, if $X_1, X_2, \cdots, X_n$ are independent standard Cauchy variables (i.e., $a = 0$, $b = 1$), then the sample mean is also a standard Cauchy variable. This seems to contradict the central limit theorem (see the discussion for the normal distribution) until one recalls that the central limit theorem requires a finite mean and variance.

When compared to the normal distribution about the same mode, the Cauchy distribution is seen to have longer tails. Therefore, it is important for representing distributions where you can get values far from the median; for example, if θ has a uniform distribution from $-\pi/2$ to $\pi/2$, then $\tan(\theta)$ has a Cauchy distribution.

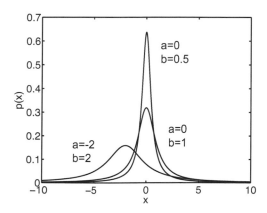

FIGURE 4.4
Probability density functions for the Cauchy distribution.

Probability density: $p(x) = \dfrac{1}{\pi b\left[1 + \frac{(x-a)^2}{b^2}\right]}$ $-\infty < x < \infty$

Range: $-\infty < x < \infty$

Parameters: $-\infty < a < \infty;$ $b > 0$

Mode: a

Median: a

r^{th} **moment about the median:** The mean and all higher moments are undefined

Characteristic function: $e^{iat - b|t|}$

Entropy: $h_X = \log_2\left(4\pi b\right)$

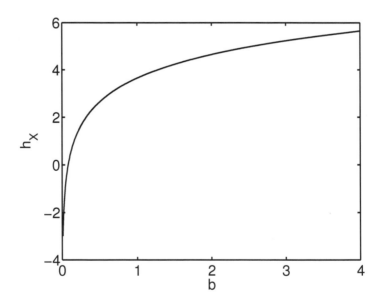

FIGURE 4.5
Differential entropy for the Cauchy distribution.

Derivation of the differential entropy for the Cauchy distribution

This is a very interesting distribution to consider, when one recalls that the moments of a Cauchy distribution are not finite. However, the differential entropy is finite and has a surprisingly simple formula. For this calculation the following integral formula will be utilized, which is taken from Gradshteyn and Ryzhik [14] formula 4.253(3), pp. 538:

$$\int_{u}^{\infty} \frac{(x-u)^{\mu-1}\ln(x)}{x^{\lambda}}dx = u^{\mu-\lambda}B(\lambda-\mu,\mu)\left[\ln(u)+\Psi(\lambda)-\Psi(\lambda-\mu)\right]$$

$$\text{if} \quad 0 < Re\{\mu\} < Re\{\lambda\}$$

where $B(\cdot,\cdot)$ is the Beta function

$$B(x,y) = \int_{0}^{1} t^{x-1}(1-t)^{y-1}dt = \frac{\Gamma(x)\Gamma(y)}{\Gamma(x+y)}$$

and $\Psi(\cdot)$ is the Psi, or Digamma function.

The differential entropy for the Cauchy distribution is given by

$$h_X = -\int_{-\infty}^{\infty} \frac{1}{\pi b\left(1+\frac{(x-a)^2}{b^2}\right)} \ln\left[\frac{1}{\pi b\left(1+\frac{(x-a)^2}{b^2}\right)}\right]dx$$

$$= -\int_{-\infty}^{\infty} \frac{1}{\pi b\left(1+\frac{(x-a)^2}{b^2}\right)} \left[\ln\left(\frac{b}{\pi}\right) - \ln\left(b^2+(x-a)^2\right)\right]dx$$

$$= -\ln\left(\frac{b}{\pi}\right)\int_{-\infty}^{\infty} \frac{1}{\pi b\left(1+\frac{(x-a)^2}{b^2}\right)}dx$$

$$+ \frac{1}{\pi b}\int_{-\infty}^{\infty} \frac{b^2}{b^2+(x-a)^2} \ln\left(b^2+(x-a)^2\right)dx$$

set $z = x - a$ in the second integral

$$= -\ln\left(\frac{b}{\pi}\right) + \frac{b}{\pi}\int_{-\infty}^{\infty} \frac{\ln(z^2+b^2)}{z^2+b^2}dz$$

$$= \ln\left(\frac{\pi}{b}\right) + \frac{2b}{\pi}\int_{0}^{\infty} \frac{\ln(z^2+b^2)}{z^2+b^2}dz$$

since the integrand is an even function; now set $y = z^2 + b^2$ in the second term, so that $dz = \frac{dy}{2\sqrt{y-b^2}}$

$$= \ln\left(\frac{\pi}{b}\right) + \frac{2b}{\pi} \int_{b^2}^{\infty} \frac{1}{y} \ln(y) \frac{1}{2\sqrt{y-b^2}} dy$$

$$= \ln\left(\frac{\pi}{b}\right) + \frac{b}{\pi} \int_{b^2}^{\infty} \frac{(y-b^2)^{-1/2} \ln(y)}{y} dy$$

But this integral is that given above with $u = b^2$, $\mu = 1/2$, $\lambda = 1$, and $0 < \mu < \lambda$

$$= \ln\left(\frac{\pi}{b}\right) + \frac{b}{\pi}(b^2)^{-1/2} B\left(\frac{1}{2}, \frac{1}{2}\right) \left[\ln(b^2) + \Psi(1) - \Psi\left(\frac{1}{2}\right)\right]$$

(Here $B(\frac{1}{2}, \frac{1}{2}) = \frac{\Gamma(\frac{1}{2})\Gamma(\frac{1}{2})}{\Gamma(1)} = \frac{\sqrt{\pi}\sqrt{\pi}}{1} = \pi$, and $\Psi(1) = -\gamma$, $\Psi(\frac{1}{2}) = -\gamma - 2\ln(2)$ where γ is the Euler-Mascheroni constant. These values of the Ψ function are given in Abramowitz and Stegun [1], p. 258; but note that we make use of the integral $\Psi(1) = \Gamma'(1) = \int_0^{\infty} \ln(t)e^{-t}dt = -\gamma$ in our calculation of the entropy for the Rayleigh distribution.)

$$= \ln\left(\frac{\pi}{b}\right) + \ln(b^2) + 2\ln(2)$$
$$= \ln(\pi) - \ln(b) + 2\ln(b) + \ln(4)$$
$$= \ln(4\pi b)$$

so

$$h_X = \log_2(4\pi b) \quad \text{bits}$$

4.3 Chi Distribution

The Chi distribution represents another important family of distributions with semi-infinite support set $[0, \infty)$. The parameter n is referred to as the "degrees of freedom." The Chi distribution arises naturally from the normal distribution since it is the distribution of the square root of the sum of squares of independent normal random variables each having zero mean and the same variance σ^2. Consequently, this distribution is widely used for error analyses.

The Chi distribution with $n = 2$ degrees of freedom is the Rayleigh distribution (with σ replaced by $\sqrt{2}\sigma$), and the Chi distribution with $n = 3$ degrees of freedom and $\sigma = \alpha\sqrt{\frac{3}{2}}$ is the Maxwell distribution.

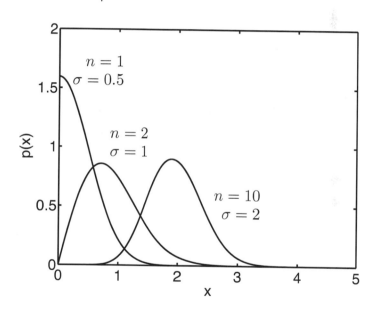

FIGURE 4.6
Probability density functions for the Chi distribution.

Probability density: $p(x) = \begin{cases} \frac{2(n/2)^{n/2}}{\sigma^n \Gamma(n/2)} x^{n-1} e^{-(n/(2\sigma^2))x^2} & : \quad 0 \le x < \infty \\ 0 & : \quad \text{otherwise} \end{cases}$

Range: $0 \le x < \infty$

Parameters: $\sigma > 0$; $\quad n$ is a positive integer

Mean: $\sigma\sqrt{\dfrac{2}{n}}\dfrac{\Gamma\left(\frac{n+1}{2}\right)}{\Gamma\left(\frac{n}{2}\right)}$

Variance: $\sigma^2\left[1-\dfrac{2}{n}\left(\dfrac{\Gamma\left(\frac{n+1}{2}\right)}{\Gamma\left(\frac{n}{2}\right)}\right)^2\right]$

r^{th} **moment about the origin:** $\sigma^r\left(\dfrac{2}{n}\right)^{r/2}\dfrac{\Gamma\left(\frac{n+r}{2}\right)}{\Gamma\left(\frac{n}{2}\right)}$

Mode: $\sqrt{\dfrac{n-1}{n}}\,\sigma$

Entropy: $h_X=\log_2\left(\dfrac{\Gamma\left(\frac{n}{2}\right)\sigma}{\sqrt{2n}}e^{[n-(n-1)\Psi(n/2)]/2}\right)$

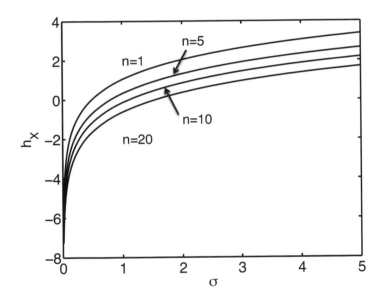

FIGURE 4.7
Differential entropy for the Chi distribution.

Derivation of the differential entropy for the Chi distribution

$$h_X = - \int_0^\infty \frac{2(n/2)^{n/2}}{\sigma^n \Gamma(n/2)} x^{n-1} e^{-(n/(2\sigma^2))x^2} \ln\left[\frac{2(n/2)^{n/2}}{\sigma^n \Gamma(n/2)} x^{n-1} e^{-(n/(2\sigma^2))x^2} \right] dx$$

$$= - \int_0^\infty \frac{2(n/2)^{n/2}}{\sigma^n \Gamma(n/2)} x^{n-1} e^{-(n/(2\sigma^2))x^2} \left[\ln\left(\frac{2(n/2)^{n/2}}{\sigma^n \Gamma(n/2)} \right) + (n-1)\ln(x) \right.$$
$$\left. - (\frac{n}{2\sigma^2})x^2 \right] dx$$

$$= - \ln\left[\frac{2(n/2)^{n/2}}{\sigma^n \Gamma(n/2)} \right] - (n-1) \frac{2(n/2)^{n/2}}{\sigma^n \Gamma(n/2)} \int_0^\infty x^{n-1} \ln(x) e^{-(n/(2\sigma^2))x^2} dx$$

$$+ \frac{2(n/2)^{n/2}}{\sigma^n \Gamma(n/2)} \frac{n}{2\sigma^2} \int_0^\infty x^{n+1} e^{-(n/(2\sigma^2))x^2} dx$$

The second of these integrals can be evaluated by means of Gradshteyn and Ryzhik [14] 3.478(1). For the first integral, let

$$t = \frac{n}{2\sigma^2} x^2$$

$$\ln(t) = \ln\left(\frac{n}{2\sigma^2} \right) + 2\ln(x)$$

$$dt = \frac{n}{\sigma^2} x \, dx$$

$$x = \left(\frac{2\sigma^2}{n} \right)^{1/2} t^{1/2}.$$

With this substitution we obtain

$$h_X = \ln\left(\frac{\sigma^n \Gamma(n/2)}{2(n/2)^{n/2}} \right)$$

$$- (n-1) \frac{2(n/2)^{n/2}}{\sigma^n \Gamma(n/2)} \int_0^\infty \left(\frac{2\sigma^2}{n} \right)^{(n-2)/2} t^{(n-2)/2} \frac{1}{2} \left(\ln(t) - \ln(n/(2\sigma^2)) \right) e^{-t} \frac{\sigma^2}{n} dt$$

$$+ \frac{2(n/2)^{n/2}}{\sigma^n \Gamma(n/2)} \frac{n}{2\sigma^2} \frac{\Gamma((n+2)/2)}{2(n/(2\sigma^2))^{(n+2)/2}}$$

$$= \ln\left(\frac{\sigma^n \Gamma(n/2)}{2(n/2)^{n/2}} \right)$$

$$- (n-1) \frac{1}{2} \frac{1}{\Gamma(n/2)} \left(\int_0^\infty t^{n/2-1} \ln(t) e^{-t} dt - \ln\left(n/(2\sigma^2) \right) \int_0^\infty t^{n/2-1} e^{-t} dt \right)$$

$$+ \frac{n}{2}$$

$$= \ln\left(\frac{\sigma^n \Gamma(n/2)}{2(n/2)^{n/2}} \right) - \frac{n-1}{2\Gamma(n/2)} \left[\Gamma'(n/2) - \ln(n/(2\sigma^2))\Gamma(n/2) \right] + \frac{n}{2}$$

where the last step follows from the formula for $\Gamma'(x)$ given in the derivation of the entropy for the Gamma distribution. Continuing the simplification

$$= n \ln(\sigma) + \ln(\Gamma(n/2)) - \ln(2) - (n/2) \ln(n/2) - \frac{n-1}{2} \Psi(n/2)$$
$$+ \frac{n-1}{2} (\ln(n/2) - 2 \ln(\sigma)) + \frac{n}{2}$$
$$= n \ln(\sigma) + \ln(\Gamma(n/2)) - \ln(2) - (n/2) \ln(n/2) - \frac{n-1}{2} \Psi(n/2)$$
$$+ \frac{n-1}{2} \ln(n/2) - (n-1) \ln(\sigma) + \frac{n}{2}$$
$$= \ln(\sigma) + \ln(\Gamma(n/2)) - \ln(2) - (1/2) \ln(n/2) - \frac{n-1}{2} \Psi(n/2) + \frac{n}{2}$$
$$= \ln \left(\frac{\sigma \Gamma(n/2)}{2\sqrt{n/2}} e^{[n-(n-1)\Psi(n/2)]/2} \right)$$
$$= \ln \left(\frac{\sigma \Gamma(n/2)}{\sqrt{2n}} e^{[n-(n-1)\Psi(n/2)]/2} \right)$$

so

$$h_X = \log_2 \left(\frac{\sigma \Gamma(n/2)}{\sqrt{2n}} e^{[n-(n-1)\Psi(n/2)]/2} \right).$$

To check this, note that if we set $n = 2$ and $\sigma = \sqrt{2}\sigma_r$ in the χ distribution, we get

$$p(x) = \frac{x}{\sigma_r^2} e^{-x^2/2\sigma_r^2}$$

which is the Rayleigh distribution. In this case the entropy becomes

$$h_X = \log_2 \left(\frac{\sqrt{2}\sigma_r}{2} e^{(2-\Psi(1))/2} \right)$$

where $\Psi(1) = \frac{\Gamma'(1)}{\Gamma(1)} = \Gamma'(1) = \int_0^\infty \ln(t) e^{-t} dt = -\gamma$ so

$$h_X = \log_2 \left(\frac{\sigma_r}{\sqrt{2}} e^{1+\gamma/2} \right)$$

which is the entropy for the Rayleigh distribution.

4.4 Chi-Squared Distribution

The Chi-squared distribution is the special case of the Gamma distribution with $\lambda = 1/2$ and $\eta = n/2$. Therefore, it has a single parameter n which is called the "degrees of freedom." The range of the random variable is from 0 to ∞. This distribution arises naturally from the normal distribution since the sum of squares of n independent normal random variables each with zero mean and unit variance has a Chi-squared distribution with n degrees of freedom.

The Chi-squared distribution occurs frequently in hypothesis testing, and is essential in the analysis of variance and in constructing confidence intervals on the standard deviation when sampling from a normal population. It is also integral to the eponymous Chi-squared Goodness-of-Fit test to determine how well a proposed theoretical distribution provides a fit to an observed histogram of data.

The sum of independent Chi-squared random variables is also Chi-squared with the degrees of freedom being the sum of the individual degrees of freedom. If X has a Chi-squared distribution, then $\sqrt{X}$ has a Chi distribution. If X and Y are independent Chi-squared random variables, then $X/(X+Y)$ has a Beta distribution.

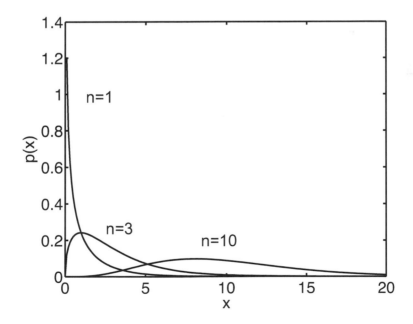

FIGURE 4.8
Probability density functions for the Chi-squared distribution.

Probability density: $p(x) = \frac{1}{2^{n/2}\Gamma(n/2)}x^{n/2-1}e^{-x/2}$

Range: $0 < x < \infty$

Parameters: n is a positive integer = degrees-of-freedom

Mean: n

Variance: $2n$

r^{th} moment about the origin: $2^r \prod\limits_{i=0}^{r-1}\left(\frac{n}{2}+i\right)$

Mode: $n - 2$ when $n \geq 2$

Characteristic function: $(1-2it)^{-n/2}$

Entropy: $h_X = \log_2\left(2\Gamma(\frac{n}{2})e^{\frac{n}{2}-(\frac{n-2}{2})\Psi(\frac{n}{2})}\right)$

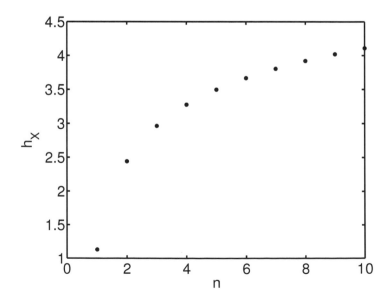

FIGURE 4.9
Differential entropy for the Chi-squared distribution.

Derivation of the differential entropy for the Chi-squared distribution

A special case of the Gamma distribution is the Chi-squared distribution, obtained by setting $\lambda = \frac{1}{2}$ and $\eta = \frac{n}{2}$, where n is a positive integer. In this case,

$$h_X = \log_2 \left(2\Gamma \left(\frac{n}{2} \right) e^{\frac{n}{2} - \left(\frac{n-2}{2} \right) \Psi \left(\frac{n}{2} \right)} \right) \qquad \text{bits}$$

4.5   Dirac Delta Distribution

The Dirac delta function $\delta(x)$ is defined to be equal to 0 for all $x \neq 0$ and to have $\int_{-\infty}^{\infty} \delta(x)dx = 1$. In fact, $\delta(x)$ is really not a function in the usual sense but it is an extremely useful concept. For example, it can be used to easily calculate Laplace transforms for a series of rectangular and triangular waves. One takes derivatives of the signal until it has been reduced to a series of weighted delta functions. Since each derivative just multiplies the Laplace transform by the complex variable s and since the Laplace transform of $\delta(x)$ is equal to 1, one just has to divide by the proper powers of s to get the transform of the original signal.

The delta function is interpreted physically as a sudden hit to a system and thus is called an "impulse function." It can be interpreted as the limit of the probability density function for the Laplace distribution as $\lambda \to \infty$. Therefore, we take the limit of the differential entropy for the Laplace distribution; that is

$$\lim_{\lambda \to \infty} \log_2 \left(\frac{2e}{\lambda} \right) = -\infty$$

as the differential entropy for $\delta(x)$.

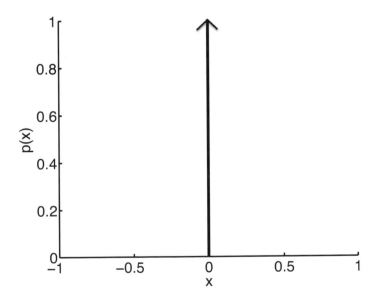

FIGURE 4.10
Probability density function for the Dirac delta distribution.

Probability density: $p(x) = \delta(x)$

Range: $-\infty < x < \infty$

Parameters: none

Mean: 0

Variance: 0

Mode: 0

Entropy: $h_X = -\infty$

4.6   Exponential Distribution

The exponential distribution is the most generally used model for time-to-failure distributions in reliability theory. The probability density function is monotonically decreasing over its semi-infinite support set. When used to represent times-to-failure, the distribution parameter λ is called the failure rate parameter. Sometimes, the model is used to represent survival times; in this case, λ is replaced by $1/\beta$ and β is called the survival parameter and is equal to the average survival duration.

In reliability studies an important function is the hazard function $haz(t)$ which calculates the probability of failure during a very small time increment given that no failure occurred previously:

$$haz(t) = \frac{p(t)}{1 - P(t)} \tag{4.2}$$

where $p(t)$ is the probability density function and $P(t)$ the cumulative distribution function for time to failure. In the case of the exponential distribution:

$$haz(t) = \frac{\lambda e^{-\lambda t}}{1 - \int_0^t \lambda e^{-\lambda y} dy} = \frac{\lambda e^{-\lambda t}}{e^{-\lambda t}} = \lambda. \tag{4.3}$$

So the exponential time-to-failure distribution is memoryless; that is, the probability of failure is constant and depends only on the length of the interval and does not depend on whether the system is in its first hour of operation or 100th hour of operation. Therefore, the time between failures is a Poisson process, that is, the failures occur continuously and independently at a constant average rate.

As a time-to-failure distribution, the exponential distribution is often a good model for electronic components, but is is even more appropriate for large complex systems. The exponential distribution is also used to represent radioactive particle decay.

Special properties include the following: If X is an exponential random variable, then so is kX. If X and Y are exponential, then $\min(X, Y)$ is exponential. If X is exponential, then e^X has a Pareto distribution. The exponential distribution is a special case of both the Gamma and Weibull distributions.

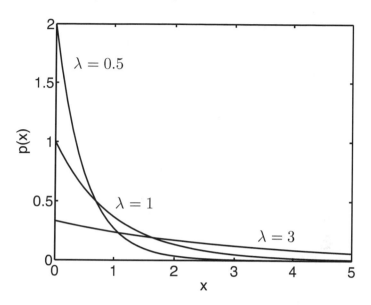

FIGURE 4.11
Probability density functions for the exponential distribution.

Probability density: $p(x) = \begin{cases} \lambda e^{-\lambda x} & : \quad x \geq 0 \\ 0 & : \quad \text{otherwise} \end{cases}$

Range: $0 \leq x < \infty$

Parameters: $\lambda > 0$

Mean: $1/\lambda$

Variance: $1/\lambda^2$

r^{th} **moment about the origin:** $\frac{r!}{\lambda^r}$

Mode: 0

Characteristic function: $\frac{\lambda}{\lambda - it}$

Entropy: $h_X = \log_2\left(\frac{e}{\lambda}\right)$

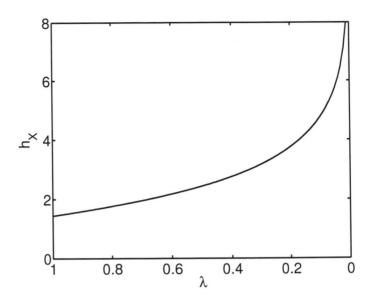

FIGURE 4.12
Differential entropy for the exponential distribution.

Derivation of the differential entropy for the exponential distribution

$$h_X = - \int_0^\infty \lambda e^{-\lambda x} \ln(\lambda e^{-\lambda x}) dx$$

$$= - \int_0^\infty \lambda e^{-\lambda x} \left(\ln(\lambda) - \lambda x \right) dx$$

$$= - \ln(\lambda) \int_0^\infty \lambda e^{-\lambda x} dx + \lambda^2 \int_0^\infty x e^{-\lambda x} dx$$

letting $y = \lambda x$ in the second integral

$$= - \ln(\lambda) + \int_0^\infty y e^{-y} dy$$

$$= - \ln(\lambda) + \Gamma(2)$$

$$= 1 - \ln(\lambda) = \ln \left(\frac{e}{\lambda} \right) \quad \text{nats}$$

where $\Gamma(x) = \int_0^\infty t^{x-1} e^{-t} dt$ is the Gamma function. In bits, the entropy for the exponential distribution becomes

$$h_X = \frac{1}{\ln(2)}\left(\ln\left(\frac{e}{\lambda}\right)\right)$$
$$= \log_2\left(\frac{e}{\lambda}\right)$$

4.7   F-Distribution

The F-distribution, also called the "Fisher-Snedecor distribution" after R. A.
Fisher and George W. Snedecor, has a probability density function with semi-
infinite support and two parameters v, w, both of which are positive integers.
As v and w become large, the probability density function approximates a
normal density function.

The F-distribution arises in the analysis of variance, since if X_1, is a Chi-
squared random variable with v degrees of freedom and X_2 is Chi-squared with
w degrees of freedom and X_1 and X_2 are independent, then $\frac{X_1/v}{X_2/w}$ follows an
F-distribution with parameters v, w. Other properties include: If X has an
F-distribution with parameters v, w, then $1/X$ has an F-distribution with
parameters w, v. Also, if X has a Student's t-distribution with parameter n,
then X^2 is an F-distributed random variable with parameters 1, n.

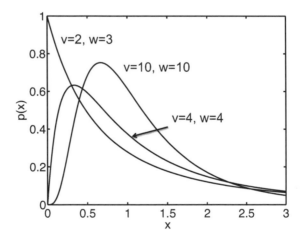

FIGURE 4.13
Probability density functions for the F-distribution.

Probability density: $p(x) = \begin{cases} \dfrac{v^{\frac{v}{2}} w^{\frac{w}{2}}}{B(\frac{v}{2},\frac{w}{2})} \dfrac{x^{\frac{v}{2}-1}}{(w+vx)^{(v+w)/2}} & : \quad 0 \le x < \infty \\ \qquad\qquad 0 & : \quad \text{otherwise} \end{cases}$

Range: $0 \le x < \infty$

Parameters: positive integers v, w

Mean: $\frac{w}{w-2}$ if $w > 2$

Variance: $\frac{2w^2(v+w-2)}{v(w-2)^2(w-4)}$ if $w > 4$

r^{th} moment about the origin: $\left(\frac{w}{v}\right)^r \frac{\Gamma(r+\frac{v}{2})\Gamma(\frac{w}{2}-r)}{\Gamma(\frac{v}{2})\Gamma(\frac{w}{2})}$ for $r < \frac{w}{2}$.

Mode: $\frac{w(v-2)}{v(w+2)}$, if $v > 1$

Entropy: $h_X = \log_2\left(\frac{w}{v}B(\frac{v}{2},\frac{w}{2})\frac{e^{(1-\frac{v}{2})\Psi(\frac{v}{2})}e^{(\frac{v+w}{2})\Psi(\frac{v+w}{2})}}{e^{(1+\frac{w}{2})\Psi(\frac{w}{2})}}\right)$

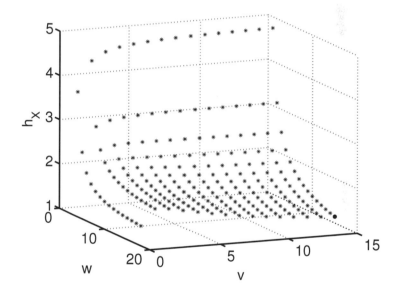

FIGURE 4.14
Differential entropy for the F-distribution.

Derivation of the differential entropy for the F-distribution

$$
\begin{aligned}
h_X &= -\int_0^\infty \frac{v^{\frac{v}{2}}w^{\frac{w}{2}}}{B(\frac{v}{2},\frac{w}{2})}\frac{x^{(\frac{v}{2})-1}}{(w+vx)^{(v+w)/2}}\ln\left[\frac{v^{\frac{v}{2}}w^{\frac{w}{2}}}{B(\frac{v}{2},\frac{w}{2})}\frac{x^{(\frac{v}{2})-1}}{(w+vx)^{(v+w)/2}}\right]dx \\
&= -\int_0^\infty \frac{v^{\frac{v}{2}}w^{\frac{w}{2}}}{B(\frac{v}{2},\frac{w}{2})}\frac{x^{(\frac{v}{2})-1}}{(w+vx)^{(v+w)/2}}\left[\ln\left(\frac{v^{\frac{v}{2}}w^{\frac{w}{2}}}{B(\frac{v}{2},\frac{w}{2})}\right)+\left(\frac{v}{2}-1\right)\ln(x)\right. \\
&\qquad\qquad \left.-\left(\frac{v+w}{2}\right)\ln(w+vx)\right]dx
\end{aligned}
$$

$$= \ln\left(\frac{B(\frac{v}{2},\frac{w}{2})}{v^{\frac{v}{2}}w^{\frac{w}{2}}}\right) - \frac{v^{\frac{v}{2}}w^{\frac{w}{2}}}{B(\frac{v}{2},\frac{w}{2})}\left(\frac{v}{2}-1\right)\int_0^\infty \frac{x^{\frac{v}{2}-1}}{(w+vx)^{(v+w)/2}}\ln(x)dx$$

$$+ \frac{v^{\frac{v}{2}}w^{\frac{w}{2}}}{B(\frac{v}{2},\frac{w}{2})}\left(\frac{v+w}{2}\right)\int_0^\infty \frac{x^{\frac{v}{2}-1}}{(w+vx)^{(v+w)/2}}\ln(w+vx)dx.$$

Here the first integral is given by:

$$\int_0^\infty \frac{x^{\frac{v}{2}-1}}{(w+vx)^{(v+w)/2}}\ln(x)dx = \frac{1}{v^{(v+w)/2}}\int_0^\infty \frac{x^{\frac{v}{2}-1}}{(x+\frac{w}{v})^{(v+w)/2}}\ln(x)dx$$

$$= \frac{1}{v^{(v+w)/2}}\left(\frac{w}{v}\right)^{-\frac{w}{2}}B\left(\frac{v}{2},\frac{w}{2}\right)\left[\ln\left(\frac{w}{v}\right)+\Psi\left(\frac{v}{2}\right)-\Psi\left(\frac{w}{2}\right)\right]$$

from Prudnikov et al. [44]. This first integral can therefore be written

$$\int_0^\infty \frac{x^{\frac{v}{2}-1}}{(w+vx)^{(v+w)/2}}\ln(x)dx = \frac{B\left(\frac{v}{2},\frac{w}{2}\right)}{v^{\frac{v}{2}}w^{\frac{w}{2}}}\left[\ln\left(\frac{w}{v}\right)+\Psi\left(\frac{v}{2}\right)-\Psi\left(\frac{w}{2}\right)\right].$$

The second integral is given by

$$\int_0^\infty \frac{x^{\frac{v}{2}-1}}{(w+vx)^{(v+w)/2}}\ln(w+vx)dx = \frac{1}{v^{(v+w)/2}}\int_0^\infty \frac{x^{\frac{v}{2}-1}}{(\frac{w}{v}+x)^{(v+w)/2}}\ln\left(v\left(\frac{w}{v}+x\right)\right)dx$$

$$= \frac{1}{v^{(v+w)/2}}\int_0^\infty \frac{x^{\frac{v}{2}-1}}{(\frac{w}{v}+x)^{(v+w)/2}}\left[\ln(v)+\ln\left(\frac{w}{v}+x\right)\right]dx$$

$$= \frac{1}{v^{(v+w)/2}}\ln(v)\int_0^\infty \frac{x^{\frac{v}{2}-1}}{(\frac{w}{v}+x)^{(v+w)/2}}dx$$

$$+ \frac{1}{v^{(v+w)/2}}\int_0^\infty \frac{x^{\frac{v}{2}-1}}{(\frac{w}{v}+x)^{(v+w)/2}}\ln\left(\frac{w}{v}+x\right)dx.$$

Rewrite the first term as:

$$\frac{1}{v^{(v+w)/2}}\ln(v)\int_0^\infty \frac{x^{\frac{v}{2}-1}}{(\frac{w}{v}+x)^{(v+w)/2}}dx = \ln(v)\int_0^\infty \frac{x^{\frac{v}{2}-1}}{(w+vx)^{(v+w)/2}}dx$$

$$= \frac{1}{w^{(v+w)/2}}\ln(v)\int_0^\infty \frac{x^{\frac{v}{2}-1}}{(1+\frac{v}{w}x)^{(v+w)/2}}dx$$

$$= \frac{1}{w^{(v+w)/2}}\ln(v)\left(\frac{w}{v}\right)^{\frac{v}{2}}B\left(\frac{v}{2},\frac{w}{2}\right)$$

(by Gradshteyn and Ryzhik [14], formula 3.194(3))

$$= \frac{\ln(v)}{w^{\frac{w}{2}}v^{\frac{v}{2}}}B\left(\frac{v}{2},\frac{w}{2}\right)$$

The second term is

$$\frac{1}{v^{(v+w)/2}}\int_0^\infty \frac{x^{\frac{v}{2}-1}}{(\frac{w}{v}+x)^{(v+w)/2}}\ln(\frac{w}{v}+x)dx$$

$$= \frac{1}{v^{(v+w)/2}}\left(\frac{w}{v}\right)^{-\frac{w}{2}}B\left(\frac{v}{2},\frac{w}{2}\right)\left[\Psi\left(\frac{v+w}{2}\right)-\Psi\left(\frac{w}{2}\right)+\ln\left(\frac{w}{v}\right)\right]$$

(by Gradshteyn and Ryzhik [14], formula 4.293(14))

$$= \frac{1}{v^{\frac{v}{2}} w^{\frac{w}{2}}} B\left(\frac{v}{2},\frac{w}{2}\right)\left[\Psi\left(\frac{v+w}{2}\right) - \Psi\left(\frac{w}{2}\right) + \ln\left(\frac{w}{v}\right)\right]$$

Therefore, the second integral is given by:

$$\int_0^\infty \frac{x^{\frac{v}{2}-1}}{(w+vx)^{(v+w)/2}} \ln(w+vx)dx$$

$$= \frac{B\left(\frac{v}{2},\frac{w}{2}\right)}{v^{\frac{v}{2}} w^{\frac{w}{2}}}\left[\ln(v) + \Psi\left(\frac{v+w}{2}\right) - \Psi\left(\frac{w}{2}\right) + \ln\left(\frac{w}{v}\right)\right]$$

$$= \frac{B\left(\frac{v}{2},\frac{w}{2}\right)}{v^{\frac{v}{2}} w^{\frac{w}{2}}}\left[\ln(w) + \Psi\left(\frac{v+w}{2}\right) - \Psi\left(\frac{w}{2}\right)\right]$$

Inserting these integral evaluations into the entropy calculation gives

$$h_X = \ln\left(\frac{B\left(\frac{v}{2},\frac{w}{2}\right)}{v^{\frac{v}{2}} w^{\frac{w}{2}}}\right) - \frac{v^{\frac{v}{2}} w^{\frac{w}{2}}}{B\left(\frac{v}{2},\frac{w}{2}\right)}\left(\frac{v}{2}-1\right)\frac{B\left(\frac{v}{2},\frac{w}{2}\right)}{v^{\frac{v}{2}} w^{\frac{w}{2}}}\left[\ln\left(\frac{w}{v}\right) + \Psi\left(\frac{v}{2}\right) - \Psi\left(\frac{w}{2}\right)\right]$$

$$+ \frac{v^{\frac{v}{2}} w^{\frac{w}{2}}}{B\left(\frac{v}{2},\frac{w}{2}\right)}\left(\frac{v+w}{2}\right)\frac{B\left(\frac{v}{2},\frac{w}{2}\right)}{v^{\frac{v}{2}} w^{\frac{w}{2}}}\left[\ln(w) + \Psi\left(\frac{v+w}{2}\right) - \Psi\left(\frac{w}{2}\right)\right]$$

$$= \ln\left(B\left(\frac{v}{2},\frac{w}{2}\right)\right) - \frac{v}{2}\ln(v) - \frac{w}{2}\ln(w) + \frac{v}{2}\ln(v) - \frac{v}{2}\ln(w) - \frac{v}{2}\Psi\left(\frac{v}{2}\right)$$

$$+ \frac{v}{2}\Psi\left(\frac{w}{2}\right)$$

$$- \ln\left(\frac{v}{w}\right) + \Psi\left(\frac{v}{2}\right) - \Psi\left(\frac{w}{2}\right) + \frac{v}{2}\ln(w) + \frac{w}{2}\ln(w) + \frac{v+w}{2}\Psi\left(\frac{v+w}{2}\right)$$

$$- \frac{v}{2}\Psi\left(\frac{w}{2}\right) - \frac{w}{2}\Psi\left(\frac{w}{2}\right)$$

$$= \ln\left(B\left(\frac{v}{2},\frac{w}{2}\right)\right) - \ln\left(\frac{v}{w}\right) + \left(1-\frac{v}{2}\right)\Psi\left(\frac{v}{2}\right) - \left(1+\frac{w}{2}\right)\Psi\left(\frac{w}{2}\right)$$

$$+ \frac{(v+w)}{2}\Psi\left(\frac{v+w}{2}\right)$$

$$= \ln\left(\frac{w}{v}B\left(\frac{v}{2},\frac{w}{2}\right)\frac{e^{(1-\frac{v}{2})\Psi(\frac{v}{2})}e^{((v+w)/2)\Psi((v+w)/2)}}{e^{(1+\frac{w}{2})\Psi(\frac{w}{2})}}\right)$$

so

$$h_X = \log_2\left(\frac{w}{v}B\left(\frac{v}{2},\frac{w}{2}\right)\frac{e^{(1-\frac{v}{2})\Psi(\frac{v}{2})}e^{((v+w)/2)\Psi((v+w)/2)}}{e^{(1+\frac{w}{2})\Psi(\frac{w}{2})}}\right) \quad \text{bits}[1].$$

---

[1]In other references (e.g., [8]) the factor preceding the Beta function, $\frac{w}{v}$, incorrectly appears as $\frac{v}{w}$.

4.8   Gamma Distribution

The Gamma distribution is a two-parameter family of single-tailed distributions. It is very versatile because of the wide variety of shapes that it can represent. It also provides the primary distribution to model random variables with semi-infinite support set. One application is inventory and maintenance where the Gamma distribution is used to model the time required for a total of exactly η independent events to occur if events happen at a constant rate λ. In this case, where η is an integer, the distribution is often referred to as the "Erlang distribution."

The exponential ($\eta = 1$) and Chi-squared ($\lambda = 1/2, \eta = $ multiple of $1/2$) are special cases of the Gamma distribution. The Gamma distribution has the property that the sum of Gamma random variables with the same rate parameter λ is again a Gamma variable with the same λ (this is readily seen from the formula for the Gamma moment generating function, $(1 - \frac{t}{\lambda})^{-\eta}$). Also, if X is Maxwell distributed, then X^2 has a Gamma distribution.

One application of the Gamma distribution is to the solution of the Tank Counterfire Duel problem. This example will be discussed in some detail since it not only demonstrates the utility of the Gamma distribution, but also provides a case where a statistical analysis has had a significant impact on public policy. Back in the 1980s laser-guided missiles and projectiles were being developed by the U.S. Army to defeat tanks on the battlefield. The tank is a very hard target; a missile must make a direct hit on the tank to disable it. The laser-guided missiles required a laser designator located in a remote ground-forward observer position which would illuminate a target tank with a directed laser beam; the laser energy reflected from this spot on the tank enables the missile seeker to guide the missile to the target tank. (Note: Today's precision guidance systems no longer require remote designation.) Upon detecting that it is being illuminated, the tank, and other tanks in its platoon, can counter by firing their main guns in an effort to destroy the laser designator[1]. We have a classic duel, with the forward observer calling in a sequence of laser-guided missiles to disable the tanks and the tanks counter-firing upon the forward observer. This is an interesting scenario; there is a complex interplay between the laser designator on-time, the rate of fire of the laser-guided missiles, and the distribution of tank counterfire response times. Other factors include the tank aiming and range estimating errors, the level to which the forward observer is protected, missile flight times, and the degree of coordination of the tank platoon. Various field tests provided data on the speed with which a tank crew could recognize and fire upon a target which suddenly threatens

---

[1]The U.S. Congress was concerned that the tanks' counterfire would quickly destroy the laser designator, thus negating the effectiveness of the entire weapon system. Before committing funds for the laser-guided weapons program, Congress mandated that the Army conduct an in-depth analysis.

the tank. There is a significant difference between the distribution of times to fire the first round, since the tank has to slew its turret and aim its main gun, and the times between firing additional rounds, but it turns out that the times between successive fires after the first round all follow the same distribution. The model needed to include the possibility of the tanks firing any number of rounds during the designator on-time (up to the limit of the tank magazines). But the time to the $k+1$ fire is the sum of $k+1$ random variables, the first being the time to first fire, and the other k being identical random variables being the time between successive fires. But in general the addition of random variables corresponds to convolution of their probability density functions, each of which requires an integration, so this problem becomes intractable beyond about 4 rounds. The situation would be manageable if the distribution of times between fires was reproductive (a distribution is defined to be reproductive if the sum of independent random variables each with such a distribution is a random variable which again is such a distribution). The Normal distribution is, of course, reproductive but it is a two-tailed distribution which is not applicable to the single-tailed distribution of times between fires. Originally, log normal distributions were used to fit tank counterfire response times, but the log normal distribution is not reproductive. The Gamma distribution in general is not reproductive, but, as noted above, the sum of any number of identically distributed Gamma variables is again a Gamma random variable, which is exactly the case for the times between fires. Gamma distributions were successfully fit both to the time to first fire and the time between fires, requiring only a single integration to calculate the tank response times. This turned out to be the key step in developing the Tank Counterfire Duel model. Congress made a decision involving billions of dollars based on this model, which concluded that the forward observer incurred no significant increase in vulnerability due to being a laser designator. This decision was a major milestone in the continuing development of precision-guided munitions [31].

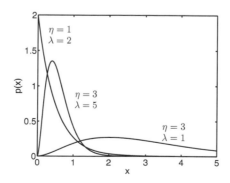

FIGURE 4.15
Probability density functions for the gamma distribution.

Probability density: $p(x) = \begin{cases} \frac{\lambda^{\eta}}{\Gamma(\eta)} x^{\eta-1} e^{-\lambda x} & : \quad 0 < x < \infty \\ 0 & : \quad \text{otherwise} \end{cases}$

Range: $0 < x < \infty$

Parameters: $\lambda > 0; \qquad \eta > 0$

Mean: η/λ

Variance: η/λ^2

r^{th} **moment about the origin:** $\frac{1}{\lambda^r} \prod\limits_{i=0}^{r-1} (\eta + i)$

Mode: $\frac{\eta-1}{\lambda}$ when $\eta \geq 1$

Characteristic function: $(1 - \frac{it}{\lambda})^{-\eta}$

Entropy: $h_X = \log_2\left(\frac{\Gamma(\eta)}{\lambda} e^{\eta + (1-\eta)\Psi(\eta)}\right)$

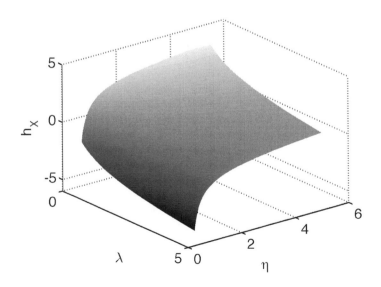

FIGURE 4.16
Differential entropy for the gamma distribution.

Derivation of the differential entropy for the gamma distribution

The Gamma function defined by

$$\Gamma(x) = \int_0^\infty t^{x-1} e^{-t} dt$$

has already been used for evaluating an integral. A graph and properties of this function can be found in the *Handbook of Mathematical Functions* by Abramowitz and Stegun [1], pp. 255 ff. The derivative of this function can be found by differentiating with respect to x inside the integral sign to obtain:

$$\Gamma'(x) = \int_0^\infty t^{x-1} \ln(t) e^{-t} dt$$

Then the Psi or Digamma function is given by

$$\Psi(x) = \frac{\Gamma'(x)}{\Gamma(x)} = \frac{d}{dx} \left(\ln\left(\Gamma(x)\right)\right).$$

A graph and properties for this function can also be found in Abramowitz and Stegun [1], pp. 258 ff. This reference also provides a table for $\Gamma(x)$ and $\Psi(x)$ for $1 \leq x \leq 2$. One can compute $\Gamma(x)$ and $\Psi(x)$ for any $x > 0$ by means of the recurrence relations

$$\Gamma(x+1) = x\Gamma(x)$$

$$\Psi(x+1) = \Psi(x) + \frac{1}{x}$$

With this background, the differential entropy for the Gamma distribution can be computed

$$h_X = -\int_0^\infty \frac{\lambda^\eta}{\Gamma(\eta)} x^{\eta-1} e^{-\lambda x} \ln\left(\frac{\lambda^\eta}{\Gamma(\eta)} x^{\eta-1} e^{-\lambda x}\right) dx$$

$$= -\int_0^\infty \frac{\lambda^\eta}{\Gamma(\eta)} x^{\eta-1} e^{-\lambda x} \left[\ln\left(\frac{\lambda^\eta}{\Gamma(\eta)}\right) + (\eta-1)\ln(x) - \lambda x\right] dx$$

$$= -\ln\left(\frac{\lambda^\eta}{\Gamma(\eta)}\right) \int_0^\infty \frac{\lambda^\eta}{\Gamma(\eta)} x^{\eta-1} e^{-\lambda x} dx - (\eta-1)\frac{\lambda^\eta}{\Gamma(\eta)} \int_0^\infty x^{\eta-1} \ln(x) e^{-\lambda x} dx$$

$$+ \frac{\lambda^{\eta+1}}{\Gamma(\eta)} \int_0^\infty x^\eta e^{-\lambda x} dx$$

set $t = \lambda x$ in both the second and third integrals; so $\ln(t) = \ln(\lambda) + \ln(x)$.

$$= -\ln\left(\frac{\lambda^\eta}{\Gamma(\eta)}\right) - \frac{(\eta-1)}{\Gamma(\eta)} \int_0^\infty \lambda^\eta \frac{t^{\eta-1}}{\lambda^{\eta-1}} \left[\ln(t) - \ln(\lambda)\right] e^{-t} \frac{1}{\lambda} dt$$

$$+ \frac{1}{\Gamma(\eta)} \int_0^\infty \lambda^{\eta+1} \frac{t^\eta}{\lambda^\eta} e^{-t} \frac{1}{\lambda} dt$$

$$= -\ln\left(\frac{\lambda^\eta}{\Gamma(\eta)}\right) - \frac{(\eta-1)}{\Gamma(\eta)} \left[\int_0^\infty t^{\eta-1} \ln(t) e^{-t} dt - \ln(\lambda) \int_0^\infty t^{\eta-1} e^{-t} dt\right]$$

$$+ \frac{1}{\Gamma(\eta)} \int_0^\infty t^\eta e^{-t} dt$$

$$= -\ln\left(\frac{\lambda^\eta}{\Gamma(\eta)}\right) - \frac{(\eta-1)}{\Gamma(\eta)} \left[\Gamma'(\eta) - \ln(\lambda)\Gamma(\eta)\right] + \frac{\Gamma(\eta+1)}{\Gamma(\eta)}$$

from the above formulas for $\Gamma(x)$ and $\Gamma'(x)$

$$= \ln(\Gamma(\eta)) - \eta \ln(\lambda) + (\eta-1)\ln(\lambda) - (\eta-1)\frac{\Gamma'(\eta)}{\Gamma(\eta)} + \eta$$

$$= \ln(\Gamma(\eta)) - \ln(\lambda) + \eta + (1-\eta)\Psi(\eta)$$

$$= \ln\left(\frac{\Gamma(\eta)}{\lambda} e^{\eta+(1-\eta)\Psi(\eta)}\right)$$

so

$$h_X = \log_2\left(\frac{\Gamma(\eta)}{\lambda} e^{\eta+(1-\eta)\Psi(\eta)}\right) \quad \text{bits}$$

4.9 Generalized Beta Distribution

The Generalized Beta distribution is just the extension of the Beta distribution to an arbitrary finite support set $[a, b]$. This distribution is appropriate for modelling events which occur in an interval with a definite minimum and maximum. For example, it is used in the project management tool PERT (Program Evaluation and Reporting Technique) for task scheduling. Here an analyst provides an optimistic estimate (t_o), a pessimistic estimate (t_p) and a most likely estimate (t_m) of the time to complete a particular task. A generalized Beta distribution for time to task completion over the interval $[t_o, t_p]$ is fit by setting the mode to be t_m and the standard deviation to be $\frac{1}{6}(t_p - t_o)$.

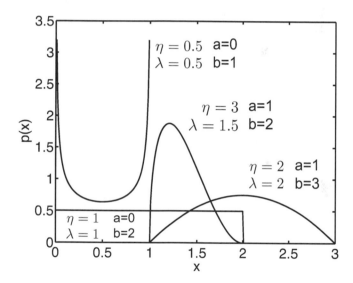

FIGURE 4.17
Probability density functions for the generalized beta distribution.

Probability density:

$$p(x) = \begin{cases} \frac{1}{b-a} \frac{\Gamma(\eta+\lambda)}{\Gamma(\eta)\Gamma(\lambda)} \left(\frac{x-a}{b-a}\right)^{\lambda-1} \left(\frac{b-x}{b-a}\right)^{\eta-1} & : \quad a \leq x \leq b \\ 0 & : \quad \text{otherwise} \end{cases}$$

Range: $a \leq x \leq b$

Parameters: $0 \leq a < b; \qquad \lambda > 0, \eta > 0$

Mean: $a + \frac{(b-a)\lambda}{\eta+\lambda}$

Variance: $\frac{(b-a)^2\eta\lambda}{(\eta+\lambda)^2(\eta+\lambda+1)}$

r^{th} **moment about the origin:** $a^r + \sum_{k=1}^{r} \binom{r}{k} (b-a)^k a^{r-k} \prod_{i=0}^{k-1} \frac{\lambda+i}{\lambda+\eta+i}$

Mode: $a + \frac{(b-a)(\lambda-1)}{\eta+\lambda+2}$ if $\lambda > 1$ and $\eta > 1$

Characteristic function: ${}_1F_1(\lambda; \lambda+\eta; it)$ (Generalized Hypergeometric series)

Entropy: $h_X = \log_2 \left(\frac{(b-a) B(\eta,\lambda) e^{(\eta+\lambda-2)\Psi(\eta+\lambda)}}{e^{(\lambda-1)\Psi(\lambda)} e^{(\eta-1)\Psi(\eta)}} \right)$

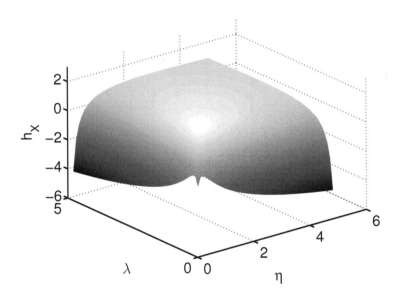

FIGURE 4.18
Differential entropy for the generalized beta distribution ($b = 5$ and $a = 2$).

Derivation of the differential entropy for the generalized beta distribution

Use the transformation $y = a + (b-a)x$ to obtain the generalized Beta distribution from the Beta distribution

$$p(y) = \begin{cases} \frac{1}{b-a} \frac{\Gamma(\eta+\lambda)}{\Gamma(\eta)\Gamma(\lambda)} \left(\frac{y-a}{b-a}\right)^{\lambda-1} \left(\frac{b-y}{b-a}\right)^{\eta-1} &: \quad a \leq y \leq b \\ 0 &: \quad \text{otherwise} \end{cases}$$

The differential entropy for the generalized Beta is now given by, using

$$B(\eta, \lambda) = \frac{\Gamma(\eta)\Gamma(\lambda)}{\Gamma(\eta + \lambda)}$$

$$h_e(Y) = -\int_a^b p(y) \ln(p(y)) dy$$

$$= -\frac{1}{B(\eta, \lambda)} \int_a^b \frac{1}{(b-a)} \left(\frac{y-a}{b-a}\right)^{\lambda-1} \left(\frac{b-y}{b-a}\right)^{\eta-1}$$

$$\times \ln\left[\frac{1}{B(\eta, \lambda)} \frac{1}{(b-a)} \left(\frac{y-a}{b-a}\right)^{\lambda-1} \left(\frac{b-y}{b-a}\right)^{\eta-1}\right] dy$$

where setting $x = \frac{y-a}{b-a}$

$$= -\frac{1}{B(\eta, \lambda)} \int_0^1 x^{\lambda-1}(1-x)^{\eta-1} \left[\ln\left(\frac{1}{b-a}\right) + \ln\left[\frac{1}{B(\eta, \lambda)} x^{\lambda-1}(1-x)^{\eta-1}\right]\right] dx$$

$$= -\ln\left(\frac{1}{b-a}\right) \left[\frac{1}{B(\eta, \lambda)} \int_0^1 x^{\lambda-1}(1-x)^{\eta-1} dx\right]$$

$$- \frac{1}{B(\eta, \lambda)} \int_0^1 x^{\lambda-1}(1-x)^{\eta-1} \ln\left[\frac{1}{B(\eta, \lambda)} x^{\lambda-1}(1-x)^{\eta-1}\right] dx$$

$$= \ln(b-a) + \text{ the Entropy of the Beta distribution}$$

$$= \ln\left(\frac{(b-a)B(\eta, \lambda)e^{(\eta+\lambda-2)\Psi(\eta+\lambda)}}{e^{(\lambda-1)\Psi(\lambda)} e^{(\eta-1)\Psi(\eta)}}\right) \text{ nats}$$

$$h_X = \log_2\left(\frac{(b-a)B(\eta, \lambda)e^{(\eta+\lambda-2)\Psi(\eta+\lambda)}}{e^{(\lambda-1)\Psi(\lambda)} e^{(\eta-1)\Psi(\eta)}}\right) \text{ bits}$$

4.10 Generalized Normal Distribution

The Generalized Normal distribution provides a parametric family of distributions which include the normal distribution when $\beta = 2$ and the Laplace distribution when $\beta = 1$. All of the distributions are symmetric but their tails behave differently than the normal distribution, which is the main reason they are used. Note that as β gets large, the distribution flattens out, resulting in a cusp at μ. In the limit as $\beta \to \infty$, the distribution becomes uniform with support set $[\mu - \alpha, \ \mu + \alpha]$.

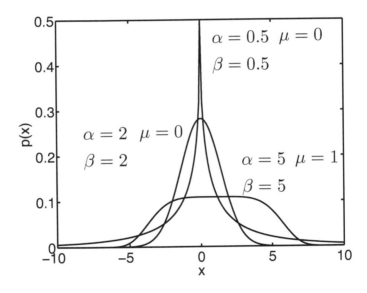

FIGURE 4.19
Probability density functions for the generalized normal distribution.

Probability Density: $p(x) = \frac{\beta}{2\alpha\Gamma(\frac{1}{\beta})}e^{-(|x-\mu|/\alpha)^{\beta}}$ $\qquad -\infty < x < \infty$

Range: $-\infty < x < \infty$

Parameters: $-\infty < \mu < \infty; \ \alpha > 0, \quad \beta > 0$

Mean: μ

Variance: $\alpha^2 \dfrac{\Gamma(\frac{3}{\beta})}{\Gamma(\frac{1}{\beta})}$

r^{th} **moment about the mean:** $= \begin{cases} 0 & : \quad r \quad \text{odd} \\ \alpha^r \dfrac{\Gamma\left(\frac{r+1}{\beta}\right)}{\Gamma\left(\frac{1}{\beta}\right)} & : \quad r \quad \text{even} \end{cases}$

r^{th} **moment about the origin:** $\mu^r + \displaystyle\sum_{\substack{k=2 \\ k \text{ even}}}^{r} \binom{r}{k} \mu^{r-k} \alpha^k \dfrac{\Gamma\left(\frac{k+1}{\beta}\right)}{\Gamma\left(\frac{1}{\beta}\right)}$

Mode: μ

Entropy: $\log_2\left(\dfrac{2\alpha\Gamma\left(\frac{1}{\beta}\right)}{\beta}e^{\frac{1}{\beta}}\right)$

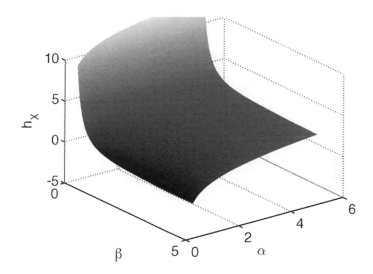

FIGURE 4.20
Differential entropy for the generalized normal distribution.

Derivation of the differential entropy for the generalized normal distribution

$$h_X = -\int_{-\infty}^{\infty} \frac{\beta}{2\alpha\Gamma\left(\frac{1}{\beta}\right)} e^{-(|x-\mu|/\alpha)^\beta} \ln\left[\frac{\beta}{2\alpha\Gamma\left(\frac{1}{\beta}\right)} e^{-(|x-\mu|/\alpha)^\beta}\right] dx$$

$$= -\int_{-\infty}^{\infty} \frac{\beta}{2\alpha\Gamma\left(\frac{1}{\beta}\right)} e^{-(|x-\mu|/\alpha)^\beta} \left[\ln\left(\frac{\beta}{2\alpha\Gamma\left(\frac{1}{\beta}\right)}\right) - \left(\frac{|x-\mu|}{\alpha}\right)^\beta\right] dx$$

$$= \ln\left(\frac{2\alpha\Gamma\left(\frac{1}{\beta}\right)}{\beta}\right) + \frac{\beta}{2\alpha\Gamma\left(\frac{1}{\beta}\right)} \int_{-\infty}^{\infty} \left(\frac{|x-\mu|}{\alpha}\right)^\beta e^{-(|x-\mu|/\alpha)^\beta} dx$$

letting $y = \frac{x-\mu}{\alpha}$

$$= \ln\left(\frac{2\alpha\Gamma\left(\frac{1}{\beta}\right)}{\beta}\right) + \frac{\beta}{2\alpha\Gamma\left(\frac{1}{\beta}\right)} \int_{-\infty}^{\infty} (|y|)^\beta e^{-(|y|)^\beta} dy$$

$$= \ln\left(\frac{2\alpha\Gamma\left(\frac{1}{\beta}\right)}{\beta}\right) + \frac{2\beta}{2\Gamma\left(\frac{1}{\beta}\right)} \int_0^\infty y^\beta e^{-y^\beta} dy$$

since the integrand is an even function

$$= \ln\left(\frac{2\alpha\Gamma\left(\frac{1}{\beta}\right)}{\beta}\right) + \frac{\beta}{\Gamma\left(\frac{1}{\beta}\right)} \left(\frac{1}{\beta}\Gamma\left(\frac{1+\beta}{\beta}\right)\right)$$

by means of Gradshteyn and Ryzhik [14] formula 3.478(1), p. 342. Since $\Gamma\left(\frac{1}{\beta}+1\right) = \frac{1}{\beta}\Gamma\left(\frac{1}{\beta}\right)$, we have

$$h_X = \ln\left(\frac{2\alpha\Gamma\left(\frac{1}{\beta}\right)}{\beta}\right) + \frac{1}{\beta}$$

$$= \ln\left(\frac{2\alpha\Gamma\left(\frac{1}{\beta}\right)}{\beta} e^{\frac{1}{\beta}}\right)$$

and

$$h_X = \log_2\left(\frac{2\alpha\Gamma\left(\frac{1}{\beta}\right)}{\beta} e^{\frac{1}{\beta}}\right) \quad \text{bits}$$

4.11 Kumaraswamy Distribution

The Kumaraswamy distribution, proposed by Poondi Kumaraswamy, is a two-parameter family of distributions with the finite support set $[0, 1]$. The formula for its probability density function is very similar to that of the Beta distribution, but the Kumaraswamy distribution is sometimes preferred because its cumulative distribution function has the closed form:

$$P(x) = 1 - (1 - x^a)^b$$

which is easier to use than the incomplete Beta function.

The Kumaraswamy distribution bears the following relationship to the Beta distribution. If X is Kumaraswamy random variable with parameters a and b, then $X = Y^{1/a}$ where Y is a Beta random variable with parameters $\lambda = 1$ and $\eta = b$.

The Kumaraswamy distribution can be extended to an arbitrary finite support set.

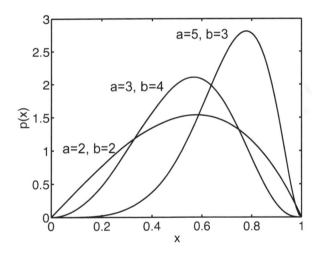

FIGURE 4.21
Probability density functions for the Kumaraswamy distribution.

Probability density: $p(x) = \begin{cases} abx^{a-1}(1 - x^a)^{b-1} & : & 0 \le x \le 1 \\ 0 & : & \text{otherwise} \end{cases}$

Range: $0 \le x \le 1$

Parameters: $a > 0, \quad b > 0$

Mean: $bB\left(1+\frac{1}{a},b\right)$

Variance: $bB\left(1+\frac{2}{a},b\right)-\left[bB\left(1+\frac{1}{a},b\right)\right]^2$

r^{th} **moment about the origin:** $bB\left(\frac{r}{a}+1,b\right)$

Mode: $\left(\frac{a-1}{ab-1}\right)^{\frac{1}{a}}$

Entropy: $h_X=\log_2\left(\frac{1}{ab}e^{\frac{b-1}{b}+\frac{a-1}{a}\left(\gamma+\Psi(b)+\frac{1}{b}\right)}\right)$

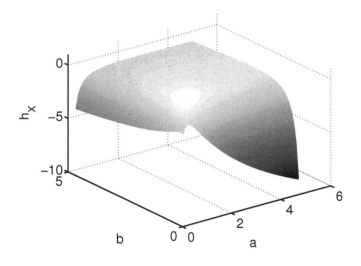

FIGURE 4.22
Differential entropy for the Kumaraswamy distribution.

Derivation of the differential entropy for the Kumaraswamy distribution

$$h_X=-\int_0^1 abx^{a-1}(1-x^a)^{b-1}\ln\left[abx^{a-1}(1-x^a)^{b-1}\right]dx$$

$$=-\int_0^1 abx^{a-1}(1-x^a)^{b-1}\left[\ln(ab)+(a-1)\ln(x)+(b-1)\ln(1-x^a)\right]dx$$

$$= -\ln(ab) - ab(a-1) \int_0^1 x^{a-1}(1-x^a)^{b-1}\ln(x)dx - ab(b-1)$$

$$\times \int_0^1 x^{a-1}(1-x^a)^{b-1}\ln(1-x^a)dx$$

The first of these integrals can be evaluated using Gradshteyn & Ryzhik [14], formula 4.253 (1) (see the derivation of the entropy for the Beta distribution). For the second integral, set $y = 1 - x^a$, $dy = -ax^{a-1}dx$.

$$= -\ln(ab) - ab(a-1)\frac{1}{a^2}B(1,b)\left[\Psi(1) - \Psi(1+b)\right]$$

$$- ab(b-1)\int_0^1 \left(\frac{1}{a}\right)y^{b-1}\ln(y)dy$$

$$= -\ln(ab) - \left(\frac{a-1}{a}\right)(\Psi(1) - \Psi(1+b)) - b(b-1)\int_0^1 y^{b-1}\ln(y)dy$$

since $B(1,b) = \frac{\Gamma(1)\Gamma(b)}{\Gamma(b+1)} = \frac{\Gamma(b)}{b\Gamma(b)} = \frac{1}{b}$, the final integral can be integrated by parts with $u = \ln(y)$, $dv = y^{b-1}dy$, $du = \frac{1}{y}dy$, $v = \frac{1}{b}y^b$

$$= -\ln(ab) - \left(\frac{a-1}{a}\right)(\Psi(1) - \Psi(1+b)) - b(b-1)\left[\frac{1}{b}y^b\ln(y)\Big|_0^1 - \frac{1}{b}\int_0^1 y^{b-1}dy\right]$$

$$= -\ln(ab) - \left(\frac{a-1}{a}\right)(\gamma - \Psi(1+b)) - b(b-1)\left[0 - \frac{1}{b^2}\right]$$

$$= -\ln(ab) - \left(\frac{a-1}{a}\right)\left(\gamma - \Psi(b) + \frac{1}{b}\right) + \frac{b-1}{b}$$

since $\Psi(b+1) = \Psi(b) + \frac{1}{b}$

$$= \ln\left(\frac{1}{ab}e^{\frac{b-1}{b} + \frac{a-1}{a}\left(\gamma + \Psi(b) + \frac{1}{b}\right)}\right)$$

where γ is Euler's constant so

$$h_X = \log_2\left(\frac{1}{ab}e^{\frac{b-1}{b} + \frac{a-1}{a}\left(\gamma + \Psi(b) + \frac{1}{b}\right)}\right) \quad \text{bits}$$

4.12   Laplace Distribution

The Laplace distribution, attributed to Pierre-Simon Laplace, has a probability density function which consists of two exponential functions: one monotonically increasing from $-\infty$ to 0, and the other monotonically decreasing from 0 to ∞. The density function resembles the normal density function but the exponent measures the absolute value of the distance from the mean at the origin rather than the square of this distance as in the normal distribution. Consequently, this distribution has larger tails than the normal distribution.

Special properties include:

- If X is a Laplace random variable, then $|X|$ has an exponential distribution

- If X and Y are exponential random variables with the same mean, then $X - Y$ is Laplace distributed

- If X and Y are Laplace random variables, then $|X/Y|$ has an F distribution

- If X and Y are Uniform over $[0, 1]$, then $\ln(X/Y)$ is Laplace distributed

The Laplace distribution is used in studies of Brownian motion.

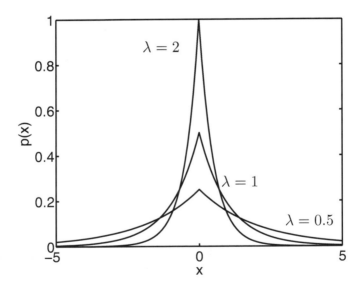

FIGURE 4.23
Probability density functions for the Laplace distribution.

Probability density: $p(x) = \frac{1}{2}\lambda e^{-\lambda|x|} \quad -\infty < x < \infty$

Range: $-\infty < x < \infty$

Parameters: $\lambda > 0$

Mean: 0

Variance: $\frac{2}{\lambda^2}$

r^{th} **moment about the origin:** $= \begin{cases} 0 & : \quad r \quad \text{odd} \\ \frac{r!}{\lambda^r} & : \quad r \quad \text{even} \end{cases}$

Mode: 0

Characteristic function: $\frac{\lambda^2}{\lambda^2 + t^2}$

Entropy: $h_X = \log_2\left(\frac{2e}{\lambda}\right)$

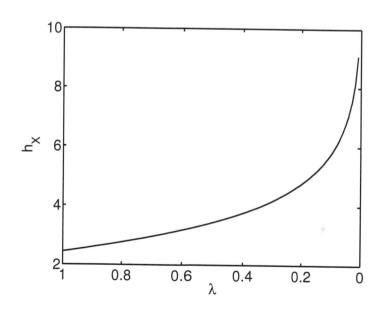

FIGURE 4.24
Differential entropy for the Laplace distribution.

Derivation of the differential entropy for the Laplace distribution

$$h_X = - \int_{-\infty}^{\infty} \frac{1}{2} \lambda e^{-\lambda |x|} \ln \left(\frac{1}{2} \lambda e^{-\lambda |x|} \right) dx$$

$$= -2 \int_{0}^{\infty} \frac{1}{2} \lambda e^{-\lambda x} \left(\ln \left(\frac{\lambda}{2} \right) - \lambda x \right) dx$$

$$= - \ln \left(\frac{\lambda}{2} \right) \int_{0}^{\infty} \lambda e^{-\lambda x} dx + \lambda^2 \int_{0}^{\infty} x e^{-\lambda x} dx$$

$$= - \ln \left(\frac{\lambda}{2} \right) + \Gamma(2)$$

$$= 1 - \ln \left(\frac{\lambda}{2} \right) = \ln \left(\frac{2e}{\lambda} \right) \quad \text{nats}$$

so

$$h_X = \log_2 \left(\frac{2e}{\lambda} \right) \quad \text{bits}$$

4.13   Log-Normal Distribution

The log-normal distribution is single-tailed with a semi-infinite range; that is, the random variable takes on only positive values. It is a very useful distribution for representing variables which are the multiplicative product of many positive-valued independent factors. This follows from the central limit theorem since X being a log-normal random variable corresponds to $\ln(X)$ being a normal random variable.

The log-normal distribution arises in reliability studies since the time to repair a system is generally log-normal. It is also often used for physiological measurements of humans and animals. Interestingly, the personal income of the 98% of the population (excluding the very wealthy 2%) is well represented by a log normal density function.

The log-normal distribution has some nice properties. The product of independent log-normal variables is again log-normal. Also, if X is a log-normal random variable, then so are aX, $1/X$, and X^a for $a \neq 0$.

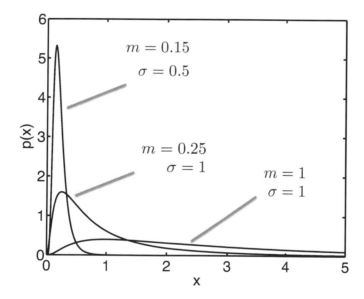

FIGURE 4.25
Probability density functions for the log-normal distributions.

$$\textbf{Probability density: } p(x) = \begin{cases} \frac{1}{\sqrt{2\pi}\sigma x}e^{-(\ln(x)-\ln(m))^2/2\sigma^2} & : \quad 0 < x < \infty \\ 0 & : \quad \text{otherwise} \end{cases}$$

Range: $0 < x < \infty$

Parameters: $m > 0;$ $\sigma > 0$

Mean: $me^{\sigma^2/2}$

Variance: $m^2 e^{\sigma^2}(e^{\sigma^2} - 1)$

r^{th} **moment about the origin:** $m^r e^{r^2 \sigma^2/2}$

Mode: $me^{-\sigma^2}$

Entropy: $h_X = \frac{1}{2}\log_2(2\pi e\sigma^2 m^2)$

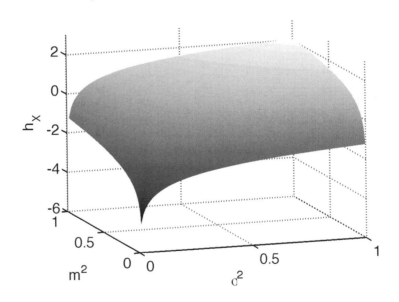

FIGURE 4.26
Differential entropy for the log-normal distribution.

Derivation of the differential entropy for the log-normal distribution

$$h_X = -\int_0^\infty \frac{1}{\sqrt{2\pi}\sigma x}e^{-(\ln(x)-\ln(m))^2/2\sigma^2}\ln\left(\frac{1}{\sqrt{2\pi}\sigma x}e^{-(\ln(x)-\ln(m))^2/2\sigma^2}\right)dx$$

$$= -\int_0^\infty \frac{1}{\sqrt{2\pi}\sigma x}e^{-(\ln(x)-\ln(m))^2/2\sigma^2}\left[\ln\left(\frac{1}{\sqrt{2\pi}\sigma}\right) - \ln(x) - \frac{1}{2\sigma^2}\left(\ln(x)-\ln(m)\right)^2\right]dx$$

$$= \ln\left(\sqrt{2\pi}\sigma\right)\int_0^\infty \frac{1}{\sqrt{2\pi}\sigma x}e^{-(\ln(x)-\ln(m))^2/2\sigma^2}dx$$

$$+ \frac{1}{\sqrt{2\pi}\sigma}\int_0^\infty \frac{\ln(x)}{x}e^{-(\ln(x)-\ln(m))^2/2\sigma^2}dx$$

$$+ \frac{1}{2\sqrt{2\pi}\sigma^3}\int_0^\infty \frac{(\ln(x)-\ln(m))^2}{x}e^{-(\ln(x)-\ln(m))^2/2\sigma^2}dx$$

set $y = \ln(x) - \ln(m)$ in both the second and third integrals

$$= \ln(\sqrt{2\pi}\sigma) + \frac{1}{\sqrt{2\pi}\sigma}\int_{-\infty}^\infty (y+\ln(m))e^{-y^2/2\sigma^2}dy + \frac{1}{2\sqrt{2\pi}\sigma^3}\int_{-\infty}^\infty y^2 e^{-y^2/2\sigma^2}dy$$

$$= \ln(\sqrt{2\pi}\sigma) + \frac{1}{\sqrt{2\pi}\sigma}\int_{-\infty}^\infty y e^{-y^2/2\sigma^2}dy$$

$$+ \frac{\ln(m)}{\sqrt{2\pi}\sigma}\int_{-\infty}^\infty e^{-y^2/2\sigma^2}dy + \frac{2}{2\sqrt{2\pi}\sigma^3}\int_0^\infty y^2 e^{-y^2/2\sigma^2}dy$$

$$= \ln(\sqrt{2\pi}\sigma) + 0 + \ln(m) + \frac{2}{\sqrt{\pi}}\int_0^\infty \left(\frac{y^2}{2\sigma^2}\right)e^{-y^2/2\sigma^2}\frac{1}{\sqrt{2}\sigma}dy$$

since the second term has an odd integrand and the third is a normal distribution. Now set $z = \frac{y}{\sqrt{2}\sigma}$ in the fourth term

$$= \ln(\sqrt{2\pi}\sigma) + \ln(m) + \frac{2}{\sqrt{\pi}}\int_0^\infty z^2 e^{-z^2}dx$$

$$= \ln(\sqrt{2\pi}\sigma m) + \frac{2}{\sqrt{\pi}}\frac{\sqrt{\pi}}{4}$$

by means of Korn & Korn [22], p. 331, formula 42

$$= \ln(\sqrt{2\pi}\sigma m) + \frac{1}{2} = \frac{1}{2}\ln(2\pi e\sigma^2 m^2)$$

so

$$h_X = \frac{1}{2}\log_2(2\pi e\sigma^2 m^2) \quad \text{bits}$$

It is of interest to compare this result to the differential entropy for the normal distribution.

4.14 Logistic Distribution

The probability density function for the Logistic distribution is a bell-shaped curve ranging over $-\infty$ to ∞ much like the normal distribution but with thicker tails. The probability density function can also be expressed in terms of hyperbolic functions as

$$p(x) = \frac{1}{4s}\operatorname{sech}^2\left(\frac{x-\mu}{2s}\right)$$

The distribution receives its name from its cumulative distribution function:

$$P(x) = \frac{1}{1 + e^{-(x-\mu)/s}}$$

which is the logistic function appearing in logistic regression and neural networks.

The logistic distribution is used in many different fields such as in biology for studying how different species compete in the wild and in medicine for describing how epidemics spread.

This distribution has an extremely simple formula for differential entropy, especially when expressed in nats:

$$h_X = \ln(s) + 2$$

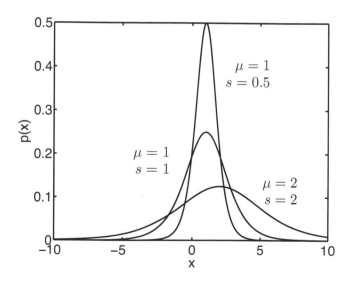

FIGURE 4.27
Probability density functions for the logistic distribution.

Probability density: $p(x) = \frac{e^{-(x-\mu)/s}}{s(1+e^{-(x-\mu)/s})^2}$

Range: $-\infty < x < \infty$

Parameters: $-\infty < \mu < \infty; \qquad s > 0$

Mean: μ

Variance: $\frac{\pi^2 s^2}{3}$

r^{th} **moment about the mean:** $= \begin{cases} 0 & : \quad r \quad \text{odd} \\ (2^r - 2)\pi^r s^r |B_r| & : \quad r \quad \text{even} \end{cases}$

where B_r is the rth Bernoulli number.

Mode: μ

Entropy: $h_X = \log_2(se^2)$

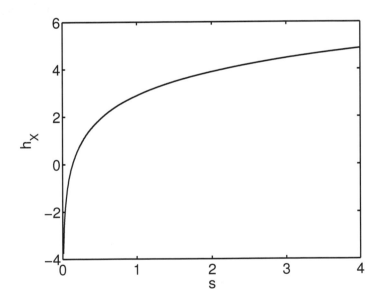

FIGURE 4.28
Differential entropy for the logistic distribution.

Derivation of the differential entropy for the logistic distribution

$$h_X = - \int_{-\infty}^{\infty} \frac{e^{-(x-\mu)/s}}{s(1 + e^{-(x-\mu)/s})^2} \ln \left(\frac{e^{-(x-\mu)/s}}{s(1 + e^{-(x-\mu)/s})^2} \right) dx.$$

Letting $y = \frac{x-\mu}{s}$

$$= - \int_{-\infty}^{\infty} \frac{e^{-y}}{s(1 + e^{-y})^2} \ln \left(\frac{e^{-y}}{s(1 + e^{-y})^2} \right) dy$$

$$= - \int_{-\infty}^{\infty} \frac{e^{-y}}{(1 + e^{-y})^2} \left[-y - \ln(s) - 2\ln(1 + e^{-y}) \right] dy$$

$$= \int_{-\infty}^{\infty} \frac{ye^{-y}}{(1 + e^{-y})^2} dy + (\ln(s)) \int_{-\infty}^{\infty} \frac{e^{-y}}{(1 + e^{-y})^2} dy + 2 \int_{-\infty}^{\infty} \frac{e^{-y}\ln(1 + e^{-y})}{(1 + e^{-y})^2} dy$$

But note that the function $\frac{e^{-y}}{(1+e^{-y})^2}$ is just the Logistic distribution with $\mu = 0$ and $s = 1$. The first integral is just the mean of this distribution and is an odd function so it is zero, and the second integral is just the integral over this same distribution, which is 1. In the third integral, let $w = e^{-y}$, so that $dw = -e^{-y}dy = -wdy$. Then

$$h_X = \ln(s) + 2 \int_0^{\infty} \frac{w\ln(1 + w)}{(1 + w)^2} \left(\frac{dw}{w} \right)$$

$$= \ln(s) + 2 \int_0^{\infty} \frac{\ln(1 + w)}{(1 + w)^2} dw$$

letting $z = 1 + w$

$$= \ln(s) + 2 \int_1^{\infty} \frac{\ln(z)}{z^2} dz$$

(Integrating by parts with $u = \ln(z)$ and $dv = \frac{1}{z^2}dz$)

$$= \ln(s) + 2 \left[-\frac{\ln(z)}{z} \Big|_1^{\infty} + \int_1^{\infty} \frac{dz}{z^2} \right]$$

$$= \ln(s) + 2 \left[0 + \int_1^{\infty} \frac{dz}{z^2} \right]$$

using L'Hopital's rule

$$= \ln(s) + 2 \left[-\frac{1}{z} \Big|_1^{\infty} \right]$$

$$= \ln(s) + 2$$

$$= \ln(se^2)$$

So

$$h_X = \log_2(se^2) \text{ bits.}$$

In the special case where $s = 1$,

$$h_X = 2 \text{ nats}$$
$$h_X = \frac{2}{\ln(2)} \text{ bits}$$

4.15   Log-Logistic Distribution

The log-logistic distribution is a two-parameter distribution with semi-infinite support whose probability density function has a shape similar to the log-normal density but with thicker tails. If X has a log-logistic distribution with parameters α, β then $\ln(X)$ has a logistic distribution with parameters $\ln(\alpha)$, β. When $\beta = 1$, the log-logistic distribution becomes a Pareto distribution if the variable is translated α units to the right.

The log-logistic distribution is particularly useful for analyses of events whose rate of occurrence increases at first then decreases, such as blood sugar glucose level before and after diagnosis and treatment for diabetes. It is also used in hydrology to represent stream flow rates and in economics to model distribution of income.

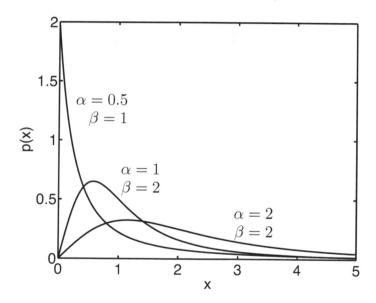

FIGURE 4.29
Probability density functions for the log-logistic distribution.

Probability density: $p(x) = \begin{cases} \dfrac{\frac{\beta}{\alpha}(\frac{x}{\alpha})^{\beta-1}}{[1+(\frac{x}{\alpha})^{\beta}]^2} & : \quad 0 \leq x < \infty \\ \qquad 0 & : \quad \text{otherwise} \end{cases}$

Range: $0 \leq x < \infty$

Parameters: $\alpha > 0; \qquad \beta > 0$

Mean: $\dfrac{\alpha\pi}{\beta \sin(\pi/\beta)}$ if $\beta > 1$

Variance: $\alpha^2 \left(\frac{2\pi}{\beta \sin(2\pi/\beta)} - \frac{\pi^2}{\beta^2 \sin^2(\pi/\beta)} \right)$ if $\beta > 2$

r^{th} **moment about the origin:** $\frac{r\pi\alpha^r}{\beta \sin(r\pi/\beta)}$ if $r < \beta$.

Mode: $\alpha \left(\frac{\beta-1}{\beta+1} \right)^{1/\beta}$ if $\beta \geq 1$, otherwise 0

Entropy: $h_X = \log_2 \left(\frac{\alpha}{\beta} e^2 \right)$

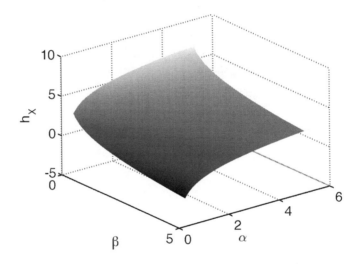

FIGURE 4.30
Differential entropy for the log-logistic distribution.

Derivation of the differential entropy for the log-logistic distribution

$$h_X = -\int_0^\infty \frac{\frac{\beta}{\alpha} \left(\frac{x}{\alpha} \right)^{\beta-1}}{[1 + \left(\frac{x}{\alpha} \right)^\beta]^2} \ln \left(\frac{\frac{\beta}{\alpha} \left(\frac{x}{\alpha} \right)^{\beta-1}}{[1 + \left(\frac{x}{\alpha} \right)^\beta]^2} \right) dx$$

$$= -\int_0^\infty \frac{\frac{\beta}{\alpha}\left(\frac{x}{\alpha} \right)^{\beta-1}}{[1 + \left(\frac{x}{\alpha} \right)^\beta]^2} \left[\ln \left(\frac{\beta}{\alpha} \right) + (\beta - 1) \ln(\frac{x}{\alpha}) - 2 \ln(1 + \left(\frac{x}{\alpha} \right)^\beta) \right] dx$$

$$= -\ln(\frac{\beta}{\alpha}) - (\beta - 1) \int_0^\infty \frac{\frac{\beta}{\alpha}(\frac{x}{\alpha})^{\beta-1}\ln(\frac{x}{\alpha})}{[1 + (\frac{x}{\alpha})^\beta]^2}dx$$

$$+ 2\int_0^\infty \frac{\frac{\beta}{\alpha}(\frac{x}{\alpha})^{\beta-1}\ln(1 + (\frac{x}{\alpha})^\beta)}{[1 + (\frac{x}{\alpha})^\beta]^2}dx$$

In each integral, set $y = (\frac{x}{\alpha})^\beta$, then $dy = \frac{\beta}{\alpha}(\frac{x}{\alpha})^{\beta-1}dx$ and $\ln(y) = \beta\ln(\frac{x}{\alpha})$. Since $\alpha > 0$ and $\beta > 0$, this becomes

$$h_X = -\ln\left(\frac{\beta}{\alpha}\right) - (\beta - 1)\frac{1}{\beta}\int_0^\infty \frac{\ln(y)}{(1+y)^2}dy + 2\int_0^\infty \frac{\ln(1+y)}{(1+y)^2}dy$$

Let's examine these two integrals. First we have:

$$\int_0^\infty \frac{ln(y)}{(1+y)^2}dy = \int_0^1 \frac{\ln(y)}{(1+y)^2}dy + \int_1^\infty \frac{\ln(y)}{(1+y)^2}dy$$

$$= -\ln(2) + \int_1^\infty \frac{\ln(y)}{(1+y)^2}dy$$

using Formula 4.231(6) of Gradshteyn and Ryzhik. The second term is integrated by parts with $y = \ln(y)$ and $dv = \frac{dy}{(1+y)^2}$). So

$$\int_0^\infty \frac{\ln(y)}{(1+y)^2}dy = -\ln(2) - \frac{\ln(y)}{1+y}\Big|_1^\infty + \int_1^\infty \frac{dy}{y(1+y)}$$

$$= -\ln(2) + 0 + \ln\left(\frac{y}{y+1}\right)\Big|_1^\infty$$

$$= -\ln(2) - \ln\left(\frac{1}{2}\right) = 0$$

The other integral, $\int_0^\infty \frac{\ln(1+y)}{(1+y)^2}dy$ was shown to be equal to 1 in the derivation of the entropy for the Logistic distribution. So

$$h_X = \ln\left(\frac{\alpha}{\beta}\right) + 2$$

$$= \ln\left(\frac{\alpha}{\beta}e^2\right)$$

and

$$h_X = \log_2\left(\frac{\alpha}{\beta}e^2\right) \qquad \text{bits}$$

A special case of this distribution when $\beta = 1$ is the Pareto distribution $p(x) = \alpha(\alpha + x)^{-2}$, $0 \le x < \infty$, or equivalently, $p(y) = \alpha y^{-2}$ $\alpha \le y < \infty$. The formula for the entropy of the Pareto distribution with $x_o = \alpha$ and $c = 1$ yields the expression $h_X = \log_2(\alpha e^2)$ as expected.

4.16   Maxwell Distribution

The Maxwell, or Maxwell-Boltzmann distribution, is named after James Clerk Maxwell and Ludwig Boltzmann whose work laid the foundation for the kinetic theory of gases. They proved that, in ideal gases with insignificant quantum effects, particle speeds and also particle energies are described by the Maxwell-Boltzmann distribution. This distribution is exactly the Chi distribution with 3 degrees of freedom with $\sigma = \sqrt{3/2}\alpha$.

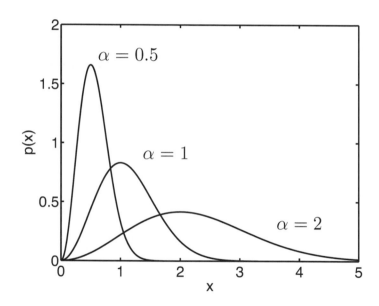

FIGURE 4.31
Probability density functions for the Maxwell distribution.

Probability density: $p(x) = \begin{cases} \frac{4}{\sqrt{\pi}} \frac{x^2 e^{-x^2/(\alpha^2)}}{\alpha^3} & : \quad 0 \le x < \infty \\ \qquad\qquad 0 & : \quad \text{otherwise} \end{cases}$

Range: $0 \le x < \infty$

Parameters: $\alpha > 0$

Mean: $\frac{2\alpha}{\sqrt{\pi}}$

Variance: $\left(\frac{3}{2} - \frac{4}{\pi}\right)\alpha^2$

r^{th} **moment about the origin:** $\frac{2}{\sqrt{\pi}}\Gamma\left(\frac{r+3}{2}\right)\alpha^r$

Mode: α

Entropy: $h_X = \log_2\left(\sqrt{\pi}\alpha e^{\gamma-1/2}\right)$

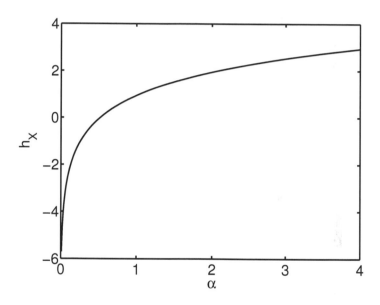

FIGURE 4.32
Differential entropy for the Maxwell distribution.

Derivation of the differential entropy for the Maxwell distribution

A special case of the χ distribution is the Maxwell distribution, obtained by setting $n = 3$ and $\sigma = \sqrt{3/2}\alpha$. In this case we have

$$h_X = \log_2\left(\frac{\Gamma(3/2)\sqrt{3/2}\alpha}{\sqrt{6}}e^{[3-2\Psi(3/2)]/2}\right).$$

Now, $\Gamma(3/2) = \frac{1}{2}\Gamma(1/2) = \frac{1}{2}\sqrt{\pi}$ and $\Psi(3/2) = \Psi(1/2) + 2 = -\gamma - 2\ln(2) + 2$
so

$$h_X = \log_2\left(\frac{\sqrt{\pi}}{4}\alpha e^{\frac{1}{2}(3+2\gamma+4\ln(2)-4)}\right)$$

$$= \log_2\left(\frac{\sqrt{\pi}}{4}\alpha e^{2\ln(2)}e^{\gamma-1/2}\right)$$

$$= \log_2\left(\sqrt{\pi}\alpha e^{\gamma-1/2}\right) \qquad \text{bits}$$

4.17 Mixed-Gaussian Distribution

The Mixed-Gaussian distribution is the only bi-modal distribution included in this text. It results from adding together two normal distributions with the same variance, σ^2, one centered at $+\mu$ and the other at $-\mu$. This produces a bimodal distribution when $\mu > \sigma\sqrt{\ln(4)}$. The moments, mode, and characteristic function for the mixed-Gaussian distribution are calculated in Appendix B.

Recently, Isserlis' Theorem has been generalized for jointly mixed-Gaussian random variables (see Section 7.6 in the Appendix), which permits the calculation of higher-order spectra in the analysis of dynamical systems (see the description for the Normal distribution for more background on this application).

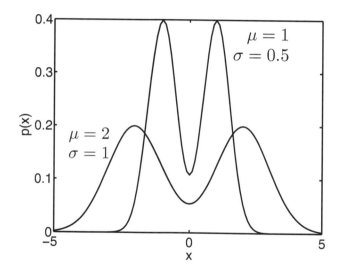

FIGURE 4.33
Probability density functions for the mixed-Gaussian distribution.

Probability density: $p(x) = \frac{1}{2\sigma\sqrt{2\pi}}\left[e^{-(x-\mu)^2/2\sigma^2} + e^{-(x+\mu)^2/2\sigma^2}\right]$

Range: $-\infty < x < \infty$

Parameters: $\mu \geq 0; \qquad \sigma > 0$

Mean: 0

Variance: $\sigma^2 + \mu^2$

$$r^{th} \text{ \bf moment about the origin:} = \begin{cases} 0 & : & r \text{ odd} \\ \mu^r + \sum\limits_{\substack{k=2 \\ k \text{ even}}}^{r} \binom{r}{k} \mu^{r-k} \dfrac{\sigma^k k!}{2^{k/2}(k/2)!} & : & r \text{ even} \end{cases}$$

$$\text{\bf Mode:} \begin{cases} 0 & : \text{ if} & \mu \leq \sigma\sqrt{\ln(4)} \\ \text{Bimodal} \quad \pm\mu & : \text{ if} & \mu > \sigma\sqrt{\ln(4)} \end{cases}$$

Characteristic function: $e^{-\sigma^2 t^2/2}\cos(\mu t)$

Entropy: $h_X = \frac{1}{2}\log_2\left(2\pi e\sigma^2\right) + L\left(\frac{\mu}{\sigma}\right)$

where $L(\cdot)$ is a monotonically increasing function of μ/σ which is 0 when $\mu/\sigma = 0$ and converges to 1 as μ/σ approaches 3.5 (This function is tabulated in Appendix 7.4).

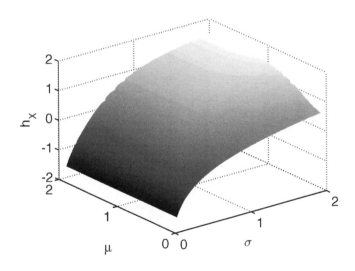

FIGURE 4.34
Differential entropy for the mixed-Gaussian distribution.

Derivation of the differential entropy for the mixed-Gaussian distribution

First, note that the probability density function can be re-written in the more compact form:

$$p(x) = \frac{1}{\sqrt{2\pi}\sigma}e^{-(x^2+\mu^2)/2\sigma^2}\cosh\left(\frac{\mu x}{\sigma^2}\right)$$

$$h_X = -\int_{-\infty}^{\infty} p(x)\ln(p(x))dx$$

$$= -\int_{-\infty}^{\infty} \frac{1}{2\sqrt{2\pi}\sigma}\left[e^{-(x-\mu)^2/2\sigma^2} + e^{-(x+\mu)^2/2\sigma^2}\right]$$

$$\times \ln\left(\frac{1}{\sqrt{2\pi}\sigma}e^{-(x^2+\mu^2)/2\sigma^2}\cosh\left(\frac{\mu x}{\sigma^2}\right)\right)dx$$

$$= -\int_{-\infty}^{\infty} \frac{1}{2\sqrt{2\pi}\sigma}\left[e^{-(x-\mu)^2/2\sigma^2} + e^{-(x+\mu)^2/2\sigma^2}\right]$$

$$\times \left[\ln(\frac{1}{\sqrt{2\pi}\sigma}) - \frac{(x^2+\mu^2)}{2\sigma^2} + \ln\left(\cosh\left(\frac{\mu x}{\sigma^2}\right)\right)\right]dx$$

$$= \ln(\sqrt{2\pi}\sigma) + \frac{\mu^2}{2\sigma^2} + \frac{1}{2\sigma^2}\int_{-\infty}^{\infty} \frac{x^2}{2\sqrt{2\pi}\sigma}(e^{-(x-\mu)^2/2\sigma^2} + e^{-(x+\mu)^2/2\sigma^2})dx$$

$$- \frac{1}{\sqrt{2\pi}\sigma}\int_{-\infty}^{\infty} e^{-(x^2+\mu^2)/2\sigma^2}\cosh\left(\frac{\mu x}{\sigma^2}\right)\ln\left(\cosh\left(\frac{\mu x}{\sigma^2}\right)\right)dx$$

$$= \ln(\sqrt{2\pi}\sigma) + \frac{\mu^2}{2\sigma^2} + \frac{1}{2\sigma^2}(\sigma^2+\mu^2)$$

$$- \frac{1}{\sqrt{2\pi}\sigma}e^{-\mu^2/2\sigma^2}\int_{-\infty}^{\infty} e^{-x^2/2\sigma^2}\cosh\left(\frac{\mu x}{\sigma^2}\right)\ln\left(\cosh\left(\mu x/\sigma^2\right)\right)dx.$$

If we let $y = \frac{\mu x}{\sigma^2}$ in this integral, the above expression becomes

$$h_X = \ln(\sqrt{2\pi}\sigma) + \frac{\mu^2}{\sigma^2} + \frac{1}{2}$$

$$- \frac{1}{\sqrt{2\pi}\sigma}e^{-\mu^2/2\sigma^2}\left(\frac{\sigma^2}{\mu}\right)\int_{-\infty}^{\infty} e^{-\sigma^2 y^2/2\mu^2}\cosh(y)\ln(\cosh(y))dy.$$

Noting that the integrand is an even function, we obtain

$$h_X = \frac{1}{2}\ln(2\pi e\sigma^2) + \frac{\mu^2}{\sigma^2}$$

$$- \frac{2}{\sqrt{2\pi}}e^{-\mu^2/2\sigma^2}\left(\frac{\sigma}{\mu}\right)\int_{0}^{\infty} e^{-\sigma^2 y^2/2\mu^2}\cosh(y)\ln(\cosh(y))dy. \qquad (4.4)$$

An analytic expression for this integral could not be found [32]. However, the first term is recognized as the entropy of a Gaussian distribution, and the remainder is tabulated in 7.4 and forms a monotonically increasing function of μ/σ which goes from 0 to $\ln(2)$. In bits, the entropy becomes

$$h_X = \frac{1}{2} \log_2 \left(2\pi e \sigma^2 \right) + L \left(\frac{\mu}{\sigma} \right)$$

where $L(\cdot)$ is the aforementioned function of μ/σ which goes from 0 to 1 and is tabulated in Appendix 7.4.

4.18   Nakagami Distribution

The Nakagami or Nakagami-m distribution is used to model signal attenuation along multiple paths. The probability density function for the Nakagami distribution has the same formula as the Chi probability density with $\sigma^2 = \Omega$ and $n = 2m$. The difference is that in the Nakagami density, m need not be an integer; in fact, m can be any real number $\geq 1/2$ (and so n is any real number ≥ 1). However, the derivation of the differential entropy proceeds exactly as that for the Chi distribution.

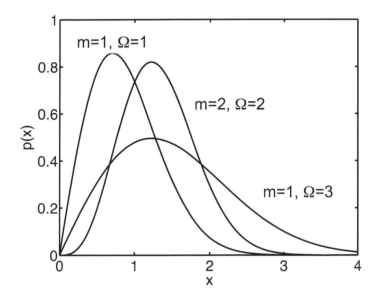

FIGURE 4.35
Probability density functions for the Nakagami distribution.

Probability density: $p(x) = \begin{cases} \frac{2}{\Gamma(m)} \left(\frac{m}{\Omega}\right)^m x^{2m-1} e^{-mx^2/\Omega} : & : \quad x > 0 \\ \qquad\qquad\qquad\qquad 0 & : \quad \text{otherwise} \end{cases}$

Range: $0 \leq x < \infty$

Parameters: $\Omega > 0$; m any real number $\geq 1/2$

Mean: $\sqrt{\frac{\Omega}{m}} \frac{\Gamma(m+1/2)}{\Gamma(m)}$

Variance: $\Omega \left[1 - \frac{1}{m} \left(\frac{\Gamma(m+1/2)}{\Gamma(m)} \right)^2 \right]$

r^{th} **moment about the origin:** $= \left(\frac{\Omega}{m} \right)^{r/2} \frac{\Gamma(m+r/2)}{\Gamma(m)}$

Mode: $\sqrt{\left(\frac{2m-1}{2m} \right) \Omega}$

Entropy: $h_X = \log_2 \left(\frac{\Gamma(m)}{2} \sqrt{\frac{\Omega}{m}} e^{[2m-(2m-1)\Psi(m)]/2} \right)$

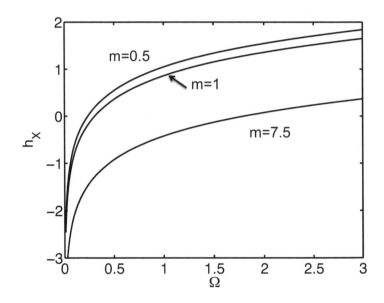

FIGURE 4.36
Differential entropy for the Nakagami distribution.

Derivation of the differential entropy for the Nakagami distribution

See derivation for the differential entropy of the Chi distribution

4.19 Normal Distribution

The Normal distribution is the most widely used probability distribution in statistics. It is often called the "Gaussian distribution" after Carl Friedrich Gauss who introduced it in 1809, although it was Pierre-Simon Laplace who developed many of its properties and proved the central limit theorem which established the fundamental role of the normal distribution in statistical analysis. Basically, the central limit theorem says that the distribution of the mean, or sum, of n independent observations taken from *any* distribution, or even n different distributions, with finite mean and variance approaches a normal distribution as n approaches infinity. So, if a random variable is the result of a large number of independent factors, we expect that distribution to be normal. For example, the normal distribution is regularly used to model electronic noise, instrumentation error, observational error in laboratory experiments, variations in temperature at specific sites, etc.

The normal distribution is also very attractive to statisticians because of the many desirable properties that it displays. Its probability density function is symmetric, bell-shaped, and two-tailed with its support set ranging from $-\infty$ to ∞. All of its cumulants beyond the first two (the mean and variance) are zero, as can be seen from the moment generating function or the characteristic function. In addition:

- Any linear combination of independent normal variables is again normal

- If X is a normal random variable, then e^X is log-normal

- If X is a normal random variable with zero mean and variance σ^2, then X^2/σ^2 is Chi-squared distributed

- If X and Y are independent normal random variables with zero means, then $\sqrt{X^2 + Y^2}$ is Rayleigh and X/Y is Cauchy distributed. Hence the log-normal, Chi-squared, Cauchy, and Rayleigh distributions are all related to the normal distribution

- Other distributions, such as the Poisson, binomial, Chi-squared, and Student's t-distribution can be approximated by the normal distribution.

- Expected values of products of jointly normal random variables can be expressed by Isserlis' Theorem (see Table of Formulas).

This last property enables the calculation of the higher-order spectra; that is, the multi-dimensional Fourier transforms of the joint cumulants, associated with the observations comprising a random process. These quantities are often used to diagnose the presence and type of nonlinearity in dynamical systems.

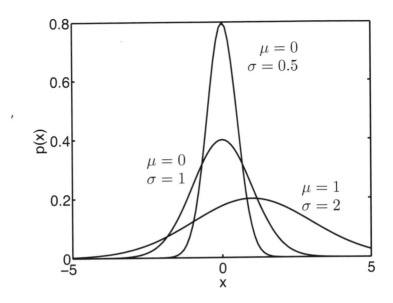

FIGURE 4.37
Probability density functions for the normal distribution.

Probability density: $p(x) = \frac{1}{\sigma\sqrt{2\pi}} e^{-\frac{1}{2\sigma^2}(x-\mu)^2}$

Range: $-\infty < x < \infty$

Parameters: $-\infty < \mu < \infty; \qquad \sigma > 0$

Mean: μ

Variance: σ^2

r^{th} **moment about the mean:** $= \begin{cases} 0 & : \quad r \quad \text{odd} \\ \frac{\sigma^r r!}{2^{r/2}(r/2)!} & : \quad r \quad \text{even} \end{cases}$

r^{th} **moment about the origin:** $\mu^r + \displaystyle\sum_{\substack{k=2 \\ k \text{ even}}}^{r} \binom{r}{k} \mu^{r-k} \frac{\sigma^k k!}{2^{k/2}(k/2)!}$

Mode: μ

Characteristic function: $e^{(i\mu t - \frac{1}{2}\sigma^2 t^2)}$

Entropy: $h_X = \frac{1}{2}\log_2(2\pi e\sigma^2)$

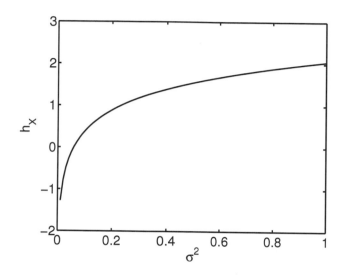

FIGURE 4.38
Differential entropy for the normal distribution.

Derivation of the differential entropy for the normal distribution

$$h_X = -\int_{-\infty}^{\infty} \frac{1}{\sigma\sqrt{2\pi}} e^{-(x-\mu)^2/2\sigma^2} \ln\left[\frac{1}{\sigma\sqrt{2\pi}} e^{-(x-\mu)^2/2\sigma^2}\right] dx$$

$$= -\int_{-\infty}^{\infty} \frac{1}{\sigma\sqrt{2\pi}} e^{-(x-\mu)^2/2\sigma^2} \left[-\ln(\sigma\sqrt{2\pi}) - \frac{(x-\mu)^2}{2\sigma^2}\right] dx$$

$$= \ln(\sigma\sqrt{2\pi}) + \frac{1}{2\sigma^2} \int_{-\infty}^{\infty} \frac{(x-\mu)^2}{\sigma\sqrt{2\pi}} e^{-(x-\mu)^2/2\sigma^2} dx$$

$$= \frac{1}{2}\ln(2\pi\sigma^2) + \frac{\sigma^2}{2\sigma^2}$$

$$= \frac{1}{2}\ln(2\pi e\sigma^2)$$

In bits, the entropy for the Normal distribution becomes

$$h_X = \frac{1}{2}\log_2\left(2\pi e\sigma^2\right)$$

4.20   Pareto Distribution

The Pareto distribution has a semi-infinite support set with starting point x_o and going to ∞. The probability density function is a simple power law with negative exponent. This distribution is due to the Italian economist Vilfredo Pareto who devised it to describe the way wealth is divided among the various levels of a society. The probability density function shows that a large fraction of the population owns a small amount of wealth and that the fraction decreases rapidly with large amounts of wealth. This is expressed in the famous Pareto "80-20 rule" which states that 20% of the population owns 80% of the wealth.

There are many examples of societal, actuarial and scientific phenomena which are aptly described by the Pareto distribution. For example, one would be the size of human population centers (many small villages, a few large cities); another would be the size of meteorites striking the earth. Still another would be the value of baseball cards; the great majority are worth less than a dollar; a few (like the famous Honus Wagner tobacco card) are worth a fortune.

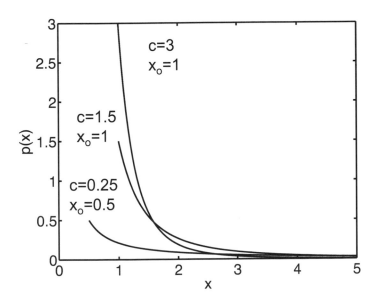

FIGURE 4.39
Probability density functions for the Pareto distribution.

$$\textbf{Probability density: } p(x) = \begin{cases} cx_o^c x^{-(c+1)} & : & x_o \leq x < \infty \\ 0 & : & \text{otherwise} \end{cases}$$

Range: $x_o \leq x < \infty$

Parameters: $c > 0$; $x_o \geq 1$

Mean: $\frac{cx_o}{c-1}$ if $c > 1$

Variance: $[\frac{c}{c-2} - \frac{c^2}{(c-1)^2}]x_o^2$ if $c > 2$

r^{th} **moment about the origin:** $\frac{c}{c-r}x_o^r$ if $c > r$

Mode: x_o

Entropy: $h_X = \log_2\left(\frac{x_o}{c}e^{1+\frac{1}{c}}\right)$

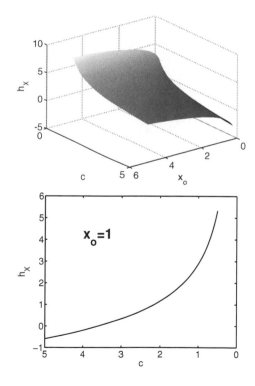

FIGURE 4.40
Differential entropy for the Pareto distribution.

Derivation of the differential entropy for the Pareto distribution

$$h_X = -\int_{x_o}^{\infty} c x_o^c x^{-(c+1)} \ln\left(c x_o^c x^{-(c+1)}\right) dx$$

$$= -\int_{x_o}^{\infty} c x_o^c x^{-(c+1)} \left[\ln(c x_o^c) - (c+1)\ln(x)\right] dx$$

$$= -\ln(c x_o^c) + c(c+1) x_o^c \int_{x_o}^{\infty} x^{-(c+1)} \ln(x) dx$$

set $y = \ln(x/x_o)$ in the integral; so $x = x_o e^y$

$$= -\ln(c x_o^c) + c(c+1) x_o^c \int_0^{\infty} x_o^{-(c+1)} e^{-(c+1)y} (y + \ln(x_o)) x_o e^y dy$$

$$= -\ln(c x_o^c) + c(c+1) \int_0^{\infty} e^{-cy} (y + \ln(x_o)) dy$$

$$= -\ln(c x_o^c) + c(c+1) \left[\int_0^{\infty} y e^{-cy} dy + \ln(x_o) \int_0^{\infty} e^{-cy} dy\right]$$

$$= -\ln(c x_o^c) + c(c+1) \left[\frac{\Gamma(2)}{c^2} + \ln(x_o)\frac{1}{c}\right]$$

where the second term was calculated as in the derivation for the exponential distribution

$$= -\ln(c) - c\ln(x_o) + \frac{c+1}{c} + (c+1)\ln(x_o)$$

$$= -\ln(c) + \ln(x_o) + \frac{c+1}{c}$$

$$= \ln\left(\frac{x_o}{c} e^{1+\frac{1}{c}}\right)$$

so

$$h_X = \log_2\left(\frac{x_o}{c} e^{1+\frac{1}{c}}\right) \quad \text{bits}$$

4.21 Rayleigh Distribution

The Rayleigh distribution, named after Lord Rayleigh, is single-tailed with the random variable always greater than or equal to zero. This distribution is applicable to the firing of artillery where one needs to know the distribution of radial errors in the plane about ground zero. In this case, projectiles are being fired at a target center where the sighting errors X and Y along perpendicular axes are independent and normally distributed with zero mean and equal variance. It turns out that the radial error $\sqrt{X^2 + Y^2}$ has a Rayleigh distribution (incidentally, the three-dimensional version of this problem leads to the Maxwell distribution).

The Rayleigh distribution is also important in statistical comunication theory, since it represents the distribution of the amplitude of the envelope of the random noise detected by a linear detector.

The Rayleigh distribution is a special case of the Chi, Weibull, and Rice distributions. If X is a Rayleigh random variable, then X^2 has a Gamma distribution.

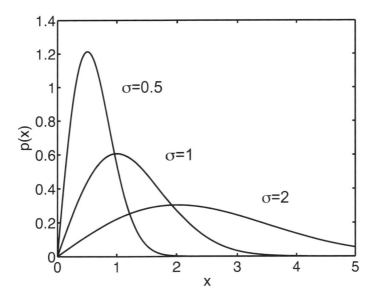

FIGURE 4.41
Probability density functions for the Rayleigh distribution.

$$\textbf{Probability density: } p(x) = \begin{cases} \frac{x}{\sigma^2} e^{-x^2/2\sigma^2} & : \quad 0 \le x < \infty \\ 0 & : \quad \text{otherwise} \end{cases}$$

Range: $0 \le x < \infty$

Parameters: $\sigma > 0$

Mean: $\sigma\sqrt{\pi/2}$

Variance: $\sigma^2(2 - \pi/2)$

r^{th} **moment about the origin:** $\sigma^r(\sqrt{2})^r\Gamma\left(1 + \frac{r}{2}\right)$

Mode: σ

Entropy: $h_X = \log_2\left(\frac{\sigma}{\sqrt{2}}e^{1+\frac{\gamma}{2}}\right)$

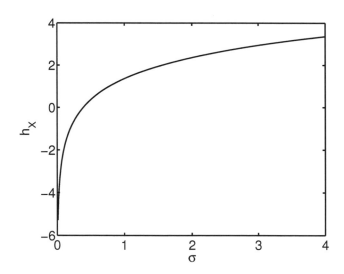

FIGURE 4.42
Differential entropy for the Rayleigh distribution.

Derivation of the differential entropy for the Rayleigh distribution

$$h_X = - \int_0^\infty \frac{x}{\sigma^2} e^{-x^2/2\sigma^2} \ln\left(\frac{x}{\sigma^2} e^{-x^2/2\sigma^2}\right) dx$$

$$= - \int_0^\infty \frac{x}{\sigma^2} e^{-x^2/2\sigma^2} \left[\ln\left(\frac{1}{\sigma^2}\right) + \ln(x) - \frac{x^2}{2\sigma^2}\right] dx$$

$$= 2\ln(\sigma) \int_0^\infty \frac{x}{\sigma^2} e^{-x^2/2\sigma^2} dx - \frac{1}{\sigma^2} \int_0^\infty x\ln(x) e^{-x^2/2\sigma^2} dx + \frac{1}{2\sigma^4} \int_0^\infty x^3 e^{-x^2/2\sigma^2} dx$$

$$= 2\ln(\sigma) - \int_0^\infty \frac{x}{\sigma^2} \ln(x) e^{-x^2/2\sigma^2} dx + \frac{1}{2\sigma^4} \frac{(2\sigma^2)^2}{2}$$

by means of Korn & Korn [22], p. 331, formula 39. Set $u = \frac{x^2}{2\sigma^2}$ in the second term; so $\ln(u) = 2\ln(x) - \ln(2\sigma^2)$.

$$= 2\ln(\sigma) + 1 - \int_0^\infty \frac{1}{2}\left(\ln(2\sigma^2) + \ln(u)\right) e^{-u} du$$

$$= 2\ln(\sigma) + 1 - \frac{1}{2}\ln(2\sigma^2) \int_0^\infty e^{-u} du - \frac{1}{2}\int_0^\infty e^{-u}\ln(u) du$$

$$= 2\ln(\sigma) + 1 - \frac{1}{2}\left(\ln(2) + 2\ln(\sigma)\right) + \frac{1}{2}\gamma$$

where the last integral is obtained by means of Korn & Korn [22], p. 332, formula 61; here γ is Euler's constant.

$$= \ln(\sigma) - \frac{1}{2}\ln(2) + 1 + \frac{1}{2}\gamma$$

$$= \ln\left(\frac{\sigma}{\sqrt{2}} e^{1+\frac{\gamma}{2}}\right)$$

so

$$h_X = \log_2\left(\frac{\sigma}{\sqrt{2}} e^{1+\frac{\gamma}{2}}\right) \quad \text{bits}$$

4.22   Rice Distribution

The Rice, or Rician distribution, named after Stephen O. Rice, has a semi-infinite support set and is a generalization of the Rayleigh distribution. it arises as the distribution of the radial error $\sqrt{X^2 + Y^2}$ when the X, Y are independent and normally distributed with the same mean and variance, but where the mean is non-zero. It also describes the distribution of the envelope of a narrowband signal in additive Gaussian noise. The probability density function shown below reduces to the Rayleigh density when $a = 0$.

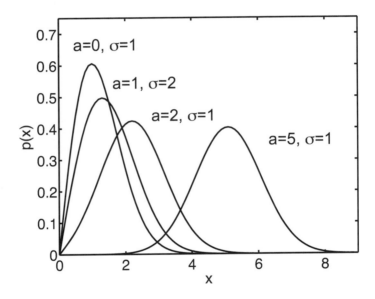

FIGURE 4.43
Probability density functions for the Rice distribution.

Probability density: $p(x) = \begin{cases} \frac{x}{\sigma^2} e^{-(x^2+a^2)/2\sigma^2} I_0\left(\frac{ax}{\sigma^2}\right) & : \quad 0 \leq x < \infty \\ 0 & : \quad \text{otherwise} \end{cases}$

where I_0 is the modified Bessel function of the first kind with order 0.
   Range: $0 \leq x < \infty$

Parameters: $a \geq 0, \ \sigma > 0$

Mean: $\sigma\sqrt{\pi/2} \, {}_1F_1\left(-\frac{1}{2}; 1; -\frac{a^2}{2\sigma^2}\right)$

where ${}_1F_1$ is the confluent hypergeometric function of the first kind, also known as Kummer's function.

Variance: $2\sigma^2 + a^2 - \frac{\pi\sigma^2}{2}\left({}_1F_1\left(-\frac{1}{2};1;-\frac{a^2}{2\sigma^2}\right)\right)^2$

r^{th} **moment about the origin:** $\sigma^r 2^{r/2}\Gamma\left(\frac{r}{2}+1\right){}_1F_1\left(-\frac{r}{2};1;-\frac{a^2}{2\sigma^2}\right)$

(Note that when r is even, the confluent hypergeometric function ${}_1F_1\left(-\frac{r}{2};1;-\frac{a^2}{2\sigma^2}\right)$ becomes the Laguerre polynomial $L_{r/2}\left(-\frac{a^2}{2\sigma^2}\right)$ and so the even moments have a much simpler form.)

Entropy: (see derivation)

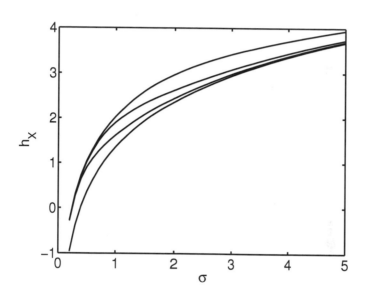

FIGURE 4.44
Differential entropy for the Rice distribution.

Derivation of the differential entropy for the Rice distribution

$$
\begin{aligned}
h_X &= -\int_0^\infty \frac{x}{\sigma^2} e^{-(x^2+a^2)/2\sigma^2} I_0\left(\frac{ax}{\sigma^2}\right) \ln\left[\frac{x}{\sigma^2} e^{-(x^2+a^2)/2\sigma^2} I_0\left(\frac{ax}{\sigma^2}\right)\right] dx \\
&= -\int_0^\infty \frac{x}{\sigma^2} e^{-(x^2+a^2)/2\sigma^2} I_0\left(\frac{ax}{\sigma^2}\right) \left[\ln\left(\frac{1}{\sigma^2}\right) + \ln(x) - \frac{x^2+a^2}{2\sigma^2} + \ln\left(I_0\left(\frac{ax}{\sigma^2}\right)\right)\right] dx
\end{aligned}
$$

$$(4.5)$$

$$= -\int_0^\infty \frac{x}{\sigma^2} e^{-(x^2+a^2)/2\sigma^2} I_0\left(\frac{ax}{\sigma^2}\right) \left[\ln\left(\frac{1}{\sigma^2}\right) - \frac{a^2}{2\sigma^2} - \frac{x^2}{2\sigma^2} + \ln(x)\right.$$

$$\left. + \ln\left(I_0\left(\frac{ax}{\sigma^2}\right)\right)\right] dx$$

$$= -\ln\left(\frac{1}{\sigma^2}\right) + \frac{a^2}{2\sigma^2} + \frac{1}{2\sigma^4} \int_0^\infty x^3 e^{-(x^2+a^2)/2\sigma^2} I_0\left(\frac{ax}{\sigma^2}\right) dx$$

$$- \frac{1}{\sigma^2} \int_0^\infty x \ln(x) e^{-(x^2+a^2)/2\sigma^2} I_0\left(\frac{ax}{\sigma^2}\right)$$

$$- \frac{1}{\sigma^2} \int_0^\infty x \ln\left(I_0\left(\frac{ax}{\sigma^2}\right)\right) e^{-(x^2+a^2)/2\sigma^2} I_0\left(\frac{ax}{\sigma^2}\right) dx$$

$$= 2\ln(\sigma) + \frac{a^2}{2\sigma^2} + \frac{1}{2\sigma^2}\mu_2' - \frac{1}{\sigma^2}\int_0^\infty x \ln(x) e^{-(x^2+a^2)/2\sigma^2} I_0\left(\frac{ax}{\sigma^2}\right) dx$$

$$- \frac{1}{\sigma^2} \int_0^\infty x e^{-(x^2+a^2)/2\sigma^2} \ln\left(I_0\left(\frac{ax}{\sigma^2}\right)\right) I_0\left(\frac{ax}{\sigma^2}\right) dx$$

(where μ_2' is the second moment about the origin)

$$= 2\ln(\sigma) + \frac{a^2}{2\sigma^2} + \frac{1}{2\sigma^2}(2\sigma^2 + a^2) - \frac{1}{\sigma^2}\int_0^\infty x \ln(x) e^{-(x^2+a^2)/2\sigma^2} I_0\left(\frac{ax}{\sigma^2}\right) dx$$

$$- \frac{1}{\sigma^2} \int_0^\infty x e^{-(x^2+a^2)/2\sigma^2} \ln\left(I_0\left(\frac{ax}{\sigma^2}\right)\right) I_0\left(\frac{ax}{\sigma^2}\right) dx$$

$$= 2\ln(\sigma) + 1 + \frac{a^2}{\sigma^2} - \frac{1}{\sigma^2}\int_0^\infty x \ln(x) e^{-(x^2+a^2)/2\sigma^2} I_0\left(\frac{ax}{\sigma^2}\right) dx$$

$$- \frac{1}{\sigma^2} \int_0^\infty x e^{-(x^2+a^2)/2\sigma^2} \ln\left(I_0\left(\frac{ax}{\sigma^2}\right)\right) I_0\left(\frac{ax}{\sigma^2}\right) dx$$

$$= 2\ln(\sigma) + 1 + \frac{a^2}{\sigma^2}$$

$$- \frac{1}{\sigma^2}\int_0^\infty x \ln(x) e^{-(x^2+a^2)/2\sigma^2} I_0\left(\frac{ax}{\sigma^2}\right) dx$$

$$- \frac{1}{\sigma^2} \int_0^\infty x e^{-(x^2+a^2)/2\sigma^2} \ln\left(I_0\left(\frac{ax}{\sigma^2}\right)\right) I_0\left(\frac{ax}{\sigma^2}\right) dx$$

Neither of these integrals can be solved in closed form. Note that in the Rayleigh case where $a = 0$, the first integral term can be evaluated (see the derivation of Differential Entropy for the Rayleigh distribution) to be $-\frac{1}{2}\ln(2) - \ln(\sigma) + \frac{\gamma}{2}$ where γ is the Euler-Mascheroni constant, and the second integral vanishes, giving the Rayleigh entropy

$$h_X = \ln(\sigma) + 1 - \ln(\sqrt{2}) + \frac{\gamma}{2} = \ln\left(\frac{\sigma}{\sqrt{2}}e^{1+\frac{\gamma}{2}}\right) \text{ nats}$$

.

So we can express the entropy for the Rice distribution in nats as

$$
h_X = \ln\left(\frac{\sigma}{\sqrt{2}}e^{1+\frac{\gamma}{2}}\right) + \frac{a^2}{\sigma^2}
$$
$$
+ \left[-\frac{1}{\sigma^2}e^{-a^2/2\sigma^2}\int_0^\infty x\ln(x)e^{-x^2/2\sigma^2}I_0\left(\frac{ax}{\sigma^2}\right)dx + \ln(\sqrt{2}) + \ln(\sigma) - \frac{\gamma}{2}\right]
$$
$$
- \frac{1}{\sigma^2}e^{-a^2/2\sigma^2}\int_0^\infty xe^{-x^2/2\sigma^2}\ln\left(I_0\left(\frac{ax}{\sigma^2}\right)\right)I_0\left(\frac{ax}{\sigma^2}\right)dx \qquad (4.6)
$$

The entropy in bits is obtained by dividing this formula by $\ln(2)$.

4.23 Simpson Distribution

The Simpson distribution, ascribed to British mathematician Thomas Simpson, is a special case of a triangular distribution. Its probability density function is symmetric about the origin over its finite range from $-a$ to a. This distribution is often used when little is known about the actual spread of values of an outcome; for example, one might want to represent a measurement error where the absolute value of the maximum error is known but little else.

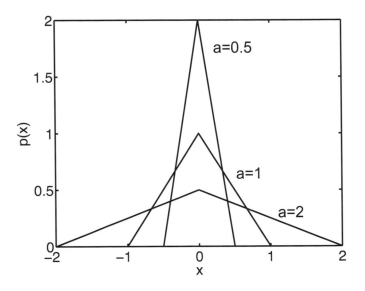

FIGURE 4.45
Probability density functions for the Simpson distribution.

Probability density: $p(x) = \begin{cases} \frac{a-|x|}{a^2} & : \quad -a \leq x \leq a \\ 0 & : \quad \text{otherwise} \end{cases}$

Range: $-a \leq x \leq a$

Parameters: $a > 0$

Mean: 0

Variance: $\frac{a^2}{6}$

r^{th} moment about the origin: $= \begin{cases} 0 & : & r & \text{odd} \\ \frac{2a^r}{(r+1)(r+2)} & : & r & \text{even} \end{cases}$

Mode: 0

Characteristic function: $\frac{2(1-\cos(at))}{a^2 t^2}$

Entropy: $h_X = \log_2(a\sqrt{e})$

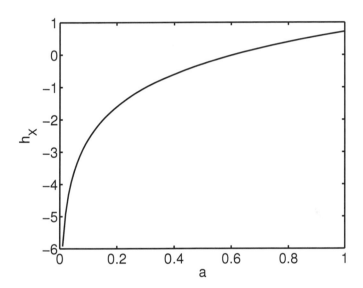

FIGURE 4.46
Differential entropy for the Simpson distribution.

Derivation of the differential entropy for the Simpson distribution

$$h_X = -\int_{-a}^{a} \left(\frac{a-|x|}{a^2}\right) \ln\left(\frac{a-|x|}{a^2}\right) dx$$

$$= -2\int_{0}^{a} \left(\frac{a-x}{a^2}\right) \ln\left(\frac{a-x}{a^2}\right) dx$$

setting $\quad y = a - x$

$$= -\frac{2}{a^2} \int_0^a y \ln\left(\frac{y}{a^2}\right) dy$$

$$= -\frac{2}{a^2} \int_0^a y \left(\ln(y) - 2\ln(a)\right) dy$$

$$= -\frac{2}{a^2} \left[\int_0^a y \ln(y) dy - 2\ln(a) \int_0^a y \, dy\right]$$

$$= -\frac{2}{a^2} \left[\left(\frac{y^2}{2}\ln(y) - \frac{y^2}{4}\right)\Big|_0^a - 2\ln(a)\left(\frac{y^2}{a}\right)\Big|_0^a\right]$$

$$= -\frac{2}{a^2} \left[\frac{a^2}{2}\ln(a) - \frac{a^2}{4} - a^2 \ln(a)\right]$$

$$= \ln(a) + \frac{1}{2} = \ln(a\sqrt{e})$$

so in bits

$$h_X = \log_2(a\sqrt{e})$$

4.24 Sine Wave Distribution

In signal processing the prototypical signal model is the sinusoid:

$$x = A\sin(\theta)$$

where θ is uniformly distributed on $[-\pi, \pi]$. (A similar derivation applies if $x = A\sin(\omega\theta)$ or $x = A\sin(\theta + \phi)$ where ω and ϕ are constants.) Restricting θ to the interval $[-\pi/2, \pi/2]$ over which x is monotonically increasing through its range of values from $-A$ to A permits the use of the usual transformation of variables technique (see Section 1.1) to calculate the probability density function of x:

$$p(x) = \frac{\frac{1}{\pi}}{\left|\frac{dx}{d\theta}\right|}$$

$$= \frac{\frac{1}{\pi}}{A\cos(\theta)}$$

$$= \frac{1}{\pi\sqrt{A^2 - x^2}} \quad \text{for} \ -A < x < A$$

which is called the "Sine Wave distribution." The calculation of the characteristic function and moments for the Sine Wave distribution is not readily available in the literature they are therefore provided in Appendix A.

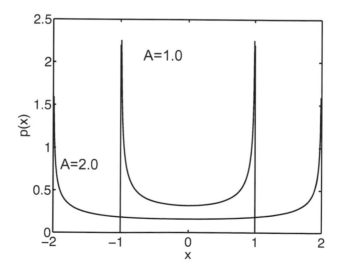

FIGURE 4.47
Probability density functions for the sine wave distribution.

Probability density: $p(x) = \begin{cases} \frac{1}{\pi\sqrt{A^2-x^2}} & : & -A < x < A \\ 0 & : & \text{otherwise} \end{cases}$

Range: $-A < x < A$

Parameters: $A > 0$

Mean: 0

Variance: $\frac{A^2}{2}$

r^{th} **moment about the origin:** $= \begin{cases} 0 & : & r \quad \text{odd} \\ A^r \left(\frac{1 \cdot 3 \cdot 5 \cdots (r-1)}{2 \cdot 4 \cdot 6 \cdots r} \right) & : & r \quad \text{even} \end{cases}$

Characteristic function: $J_o(At)$ where J_o is the 0th order Bessel function of the first kind.

Entropy: $h_X = \log_2\left(\frac{\pi A}{2}\right)$

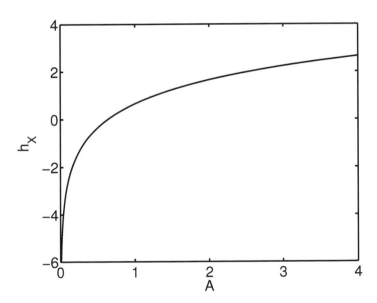

FIGURE 4.48
Differential entropy for the sine wave distribution.

Derivation of the differential entropy for the sine wave distribution

The calculation for the entropy of the sine wave $x = A\sin(\theta)$ or $x = A\sin(\omega\theta)$ or $x = A\sin(\theta + \phi)$ where θ is uniformly distributed on the interval $[-\pi, \pi]$ and ω, ϕ are constants, is given in a separate report [33]. The derivation is repeated here for completeness

$$h_X = -\int_{-A}^{A} \frac{1}{\pi\sqrt{A^2 - x^2}} \ln\left[\frac{1}{\pi\sqrt{A^2 - x^2}}\right] dx$$

$$= \frac{2}{\pi} \int_{0}^{A} \frac{1}{\sqrt{A^2 - x^2}} \ln\left[\pi\sqrt{A^2 - x^2}\right] dx$$

Making the change of variables: $w = \frac{\sqrt{A^2 - x^2}}{A}$, $dw = \frac{-x}{A\sqrt{A^2 - x^2}} dx$, and $x = A\sqrt{1 - w^2}$ leads to

$$h_X = \frac{2}{\pi} \int_{0}^{1} \ln\left[\pi A w\right] \frac{dw}{\sqrt{1 - w^2}}$$

$$= \frac{2}{\pi} \int_{0}^{1} \frac{\ln(w)dw}{\sqrt{1 - w^2}} + \frac{2}{\pi} \ln\left[\pi A\right] \int_{0}^{1} \frac{dw}{\sqrt{1 - w^2}}$$

$$= \frac{2}{\pi} \left(-\frac{\pi}{2}\ln(2)\right) + \frac{2}{\pi}\ln(\pi A)\sin^{-1}(w)\Big|_{0}^{1}$$

where the first integral is given in Gradshteyn and Ryzhik (formula 4.241(7)). The differential entropy for a sine wave is therefore

$$h_X = -\ln(2) + \ln(\pi A)$$

$$= \ln\left[\frac{\pi A}{2}\right].$$

In bits, the entropy for the sine wave distribution becomes

$$h_X = \log_2\left(\frac{\pi A}{2}\right).$$

4.25   Student's t-Distribution

This distribution was actually derived by William S. Gosset and published in 1908. Mr. Gosset worked in a brewery in Dublin which did not permit publication by its employees, so it was published under the pseudonym "Student," a name that has remained attached to Mr. Gosset's distribution ever since.

The need for this distribution arose from the statistical problem of estimating the mean of a normally distributed population when you have a relatively small sample. The procedure is as follows: Suppose independent samples $x_1, x_2, \cdots, x_n$ are drawn from a population which has a normal distribution with actual mean μ and variance σ^2. Nearly always, the true variance is unknown and must be estimated from the data. The sample mean is given by

$$\bar{x} = \frac{1}{n} \sum_{i=1}^{n} x_i$$

and the sample variance by

$$s^2 = \frac{1}{n-1} \sum_{i=1}^{n} (x_i - \bar{x})^2$$

Since $\bar{x}$ is normally distributed with mean μ and variance σ^2/n, the statistic

$$W = \frac{\bar{x} - \mu}{\sigma/\sqrt{n}}$$

is normally distributed with zero mean and variance 1. But the trouble is that you do not know σ^2 and s^2 will often be a poor estimate of σ^2 when n is small. However, the statistic

$$V = (n-1)s^2/\sigma^2$$

has a Chi-squared distribution with $(n-1)$ degrees-of-freedom, and it has been proven that the statistic

$$T = \frac{W}{\sqrt{\frac{V}{(n-1)}}} = \frac{\bar{x} - \mu}{s/\sqrt{n}}$$

has a Student's t-distribution, so this statistic can be used to test if $\bar{x}$ is a proper estimate of μ. In this same way, the Student's t-distribution is used in hypothesis testing of the significance of two different sample means and for deriving confidence levels for mean estimation.

The Student's t-distribution probability density function is bell-shaped like the normal distribution but it has larger tails. Actual data frequently include outliers that increase the size of the tails, so the Student's t-distribution provides a suitable model for such data.

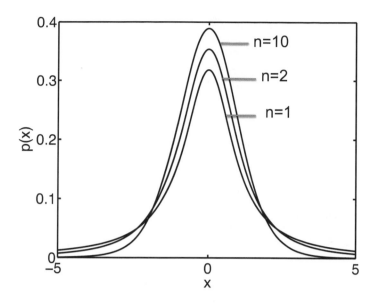

FIGURE 4.49
Probability density functions for the Student's t-distribution.

Probability density: $p(x) = \frac{1}{\sqrt{n}B(\frac{1}{2},\frac{n}{2})}(1+\frac{x^2}{n})^{-(n+1)/2}$

Range: $-\infty < x < \infty$

Parameters: n is a positive integer

Mean: 0

Variance: $\frac{n}{n-2}$ for $n > 2$

r^{th} **moment about the origin:**

$$= \begin{cases} 0 & : \quad r \quad \text{odd} \quad \text{and} \quad r < n \\ \frac{1\cdot3\cdot5\cdots(r-1)n^{r/2}}{(n-2)(n-4)\cdots(n-r)} & : \quad r \quad \text{even} \quad \text{and} \quad r < n \end{cases}$$

Mode: 0

Characteristic function: $\frac{2n^{n/4}}{\Gamma(n/2)}\left(\frac{|t|}{2}\right)^{n/2}K_{n/2}\left(\sqrt{n}|t|\right)$

where $K_{n/2}(\cdot)$ is the Bessel function of imaginary argument.

Entropy: $h_X = \log_2\left(\sqrt{n}B\left(\frac{1}{2},\frac{n}{2}\right)e^{\frac{(n+1)}{2}\left[\Psi\left(\frac{(n+1)}{2}\right)-\Psi\left(\frac{n}{2}\right)\right]}\right)$

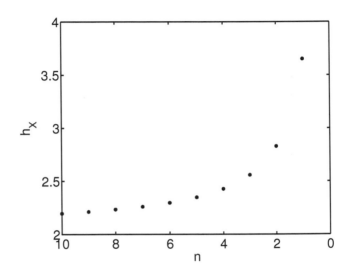

FIGURE 4.50
Differential entropy for the Student's t-distribution.

Derivation of the differential entropy for the Student's t-distribution

Noting that $B(\cdot,\cdot)$ is the Beta function defined by $B(\eta,\lambda) = \frac{\Gamma(\eta)\Gamma(\lambda)}{\Gamma(\eta+\lambda)}$, we have

$$h_X = -\int_{-\infty}^{\infty}\frac{1}{\sqrt{n}B(\frac{1}{2},\frac{n}{2})}\left(1+\frac{x^2}{n}\right)^{-(n+1)/2}\ln\left[\frac{1}{\sqrt{n}B(\frac{1}{2},\frac{n}{2})}\left(1+\frac{x^2}{n}\right)^{-(n+1)/2}\right]dx$$

$$= -\frac{1}{\sqrt{n}B(\frac{1}{2},\frac{n}{2})}\int_{-\infty}^{\infty}\left(1+\frac{x^2}{n}\right)^{-\left(\frac{n+1}{2}\right)}\left[\ln\left(\frac{1}{\sqrt{n}B(\frac{1}{2},\frac{n}{2})}\right)-\left(\frac{n+1}{2}\right)\ln\left(1+\frac{x^2}{n}\right)\right]$$

$$= \ln\left(\sqrt{n}B\left(\frac{1}{2},\frac{n}{2}\right)\right)+\frac{n+1}{2\sqrt{n}B(\frac{1}{2},\frac{n}{2})}\int_{-\infty}^{\infty}\left(1+\frac{x^2}{n}\right)^{-(n+1)/2}\ln\left(1+\frac{x^2}{n}\right)dx$$

$$= \ln\left(\sqrt{n}B\left(\frac{1}{2},\frac{n}{2}\right)\right)+\frac{n+1}{\sqrt{n}B(\frac{1}{2},\frac{n}{2})}\int_{0}^{\infty}\left(1+\frac{x^2}{n}\right)^{-(n+1)/2}\ln\left(1+\frac{x^2}{n}\right)dx.$$

since the integrand is an even function; now set $y = 1 + \frac{x^2}{n}$ in the integral, so that $x = \sqrt{n}(y-1)^{1/2}$ and $dy = \frac{2x}{n}dx = \frac{2}{\sqrt{n}}(y-1)^{1/2}dx$.

$$= \ln\left(\sqrt{n}B\left(\frac{1}{2},\frac{n}{2}\right)\right) + \frac{n+1}{2B(\frac{1}{2},\frac{n}{2})}\int_1^\infty \frac{(y-1)^{-1/2}\ln(y)}{y^{(\frac{n+1}{2})}}dy.$$

But the integral here is the same as that used in the derivation of entropy for the Cauchy distribution (Gradshteyn and Ryzhik, formula 4.253(3)) with $u = 1$, $\mu = 1/2$, and $\lambda = (n+1)/2$ and $0 < \mu < \lambda$, since n is a positive integer

$$= \ln\left(\sqrt{n}B\left(\frac{1}{2},\frac{n}{2}\right)\right) + \frac{n+1}{2B\left(\frac{1}{2},\frac{n}{2}\right)}B\left(\frac{1}{2},\frac{n}{2}\right)\left[\Psi\left(\frac{n+1}{2}\right) - \Psi\left(\frac{n}{2}\right)\right]$$

$$= \ln\left(\sqrt{n}B\left(\frac{1}{2},\frac{n}{2}\right)\right) + \frac{n+1}{2}\left[\Psi\left(\frac{n+1}{2}\right) - \Psi\left(\frac{n}{2}\right)\right]$$

$$= \ln\left(\sqrt{n}B\left(\frac{1}{2},\frac{n}{2}\right)e^{\frac{(n+1)}{2}[\Psi(\frac{n+1}{2})-\Psi(\frac{n}{2})]}\right)$$

so

$$h_X = \log_2\left(\sqrt{n}B(\frac{1}{2},\frac{n}{2})e^{\frac{(n+1)}{2}[\Psi(\frac{n+1}{2})-\Psi(\frac{n}{2})]}\right) \quad \text{bits.}$$

Note that the Student's t-distribution for $n = 1$ is the Cauchy distribution with $a = 0$ and $b = 1$. Since $B(\frac{1}{2},\frac{1}{2}) = \pi$, $\Psi(1) = -\gamma$ and $\Psi(\frac{1}{2}) = -\gamma - 2\ln(2)$, the above formula reduces to $\log_2(\pi e^{2\ln(2)}) = \log_2(4\pi)$ as expected.

4.26   Truncated Normal Distribution

The truncated Normal Distribution is obtained by taking the usual Normal distribution, truncating it on both ends and then normalizing it so that the probability density function integrates to 1. The following notation will be used in presenting the statistics of the distribution:

$$\phi(x) = \frac{1}{\sqrt{2\pi}} e^{-x^2/2} \text{ the standard normal PDF}$$

$$P_X(x) = \frac{1}{\sqrt{2\pi}} \int_{-\infty}^{x} e^{-t^2/2} dt \text{ the standard normal CDF}$$

$$Z = P_X\left(\frac{b-\mu}{\sigma}\right) - P_X\left(\frac{a-\mu}{\sigma}\right)$$

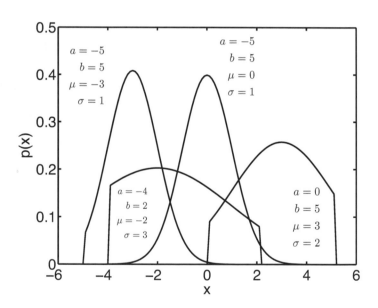

FIGURE 4.51
Probability density functions for the truncated normal distribution.

Probability density: $p(x) = \begin{cases} \frac{1}{Z\sigma\sqrt{2\pi}} e^{-(x-\mu)^2/2\sigma^2} & : \quad a \leq x \leq b \\ 0 & : \quad \text{otherwise} \end{cases}$

Range: $a \leq x \leq b$

Parameters: $a < \mu < b$, $\sigma > 0$

Mean: $\mu + \frac{\sigma}{Z}\left(\phi\left(\frac{a-\mu}{\sigma}\right) - \phi\left(\frac{b-\mu}{\sigma}\right)\right)$

Variance: $\sigma^2\left[1 + \frac{1}{Z}\left(\left(\frac{a-\mu}{\sigma}\right)\phi\left(\frac{a-\mu}{\sigma}\right) - \left(\frac{b-\mu}{\sigma}\right)\phi\left(\frac{b-\mu}{\sigma}\right)\right)\right.$
$\left. - \frac{1}{Z^2}\left(\phi\left(\frac{a-\mu}{\sigma}\right) - \phi\left(\frac{b-\mu}{\sigma}\right)\right)^2\right]$

r^{th} **moment about the origin:** $= \sum\limits_{k=0}^{r}\binom{r}{k}\tilde{\mu}_k\mu^{r-k}$

where for k even:

$$\tilde{\mu}_k = \frac{\sigma^k}{Z}\left(\left(\frac{a-\mu}{\sigma}\right)^{k-1}\phi\left(\frac{a-\mu}{\sigma}\right) - \left(\frac{b-\mu}{\sigma}\right)^{k-1}\phi\left(\frac{b-\mu}{\sigma}\right)\right)$$

$$+ \frac{\sigma^k}{Z}(k-1)\left(\left(\frac{a-\mu}{\sigma}\right)^{k-3}\phi\left(\frac{a-\mu}{\sigma}\right) - \left(\frac{b-\mu}{\sigma}\right)^{k-3}\phi\left(\frac{b-\mu}{\sigma}\right)\right)$$

$$+ \frac{\sigma^k}{Z}(k-1)(k-3)\left(\left(\frac{a-\mu}{\sigma}\right)^{k-5}\phi\left(\frac{a-\mu}{\sigma}\right) - \left(\frac{b-\mu}{\sigma}\right)^{k-5}\phi\left(\frac{b-\mu}{\sigma}\right)\right)$$

$$+ \cdots$$

$$+ \frac{\sigma^k}{Z}((k-1)(k-3)\cdots1)\left(\left(\frac{a-\mu}{\sigma}\right)\phi\left(\frac{a-\mu}{\sigma}\right) - \left(\frac{b-\mu}{\sigma}\right)\phi\left(\frac{b-\mu}{\sigma}\right)\right)$$

$$+ ((k-1)(k-3)\cdots1)\sigma^k$$

and for k odd:

$$\tilde{\mu}_k = \frac{\sigma^k}{Z}\left(\left(\frac{a-\mu}{\sigma}\right)^{k-1}\phi\left(\frac{a-\mu}{\sigma}\right) - \left(\frac{b-\mu}{\sigma}\right)^{k-1}\phi\left(\frac{b-\mu}{\sigma}\right)\right)$$

$$+ \frac{\sigma^k}{Z}(k-1)\left(\left(\frac{a-\mu}{\sigma}\right)^{k-3}\phi\left(\frac{a-\mu}{\sigma}\right) - \left(\frac{b-\mu}{\sigma}\right)^{k-3}\phi\left(\frac{b-\mu}{\sigma}\right)\right)$$

$$+ \frac{\sigma^k}{Z}(k-1)(k-3)\left(\left(\frac{a-\mu}{\sigma}\right)^{k-5}\phi\left(\frac{a-\mu}{\sigma}\right) - \left(\frac{b-\mu}{\sigma}\right)^{k-5}\phi\left(\frac{b-\mu}{\sigma}\right)\right)$$

$$+ \cdots$$

$$+ \frac{\sigma^k}{Z}((k-1)(k-3)\cdots2)\left(\phi\left(\frac{a-\mu}{\sigma}\right) - \phi\left(\frac{b-\mu}{\sigma}\right)\right)$$

Mode: μ (as long as $a < \mu < b$)

Entropy: $h_X = \log_2\left(\sqrt{2\pi e}\,\sigma Z e^{\frac{1}{2Z}\left[\left(\frac{a-\mu}{\sigma}\right)\phi\left(\frac{a-\mu}{\sigma}\right) - \left(\frac{b-\mu}{\sigma}\right)\phi\left(\frac{b-\mu}{\sigma}\right)\right]}\right)$

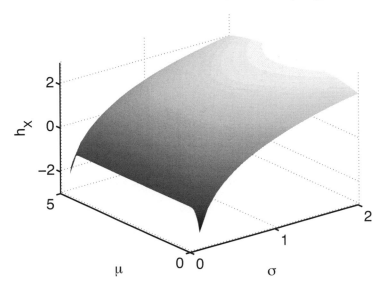

FIGURE 4.52
Differential entropy for the truncated normal distribution $(a = 0, b = 5)$.

Derivation of the differential entropy for the truncated normal distribution

We use the following notation introduced in the definition of the truncated normal probability density function:

$$\phi(x) = \frac{1}{\sqrt{2\pi}} e^{-x^2/2}$$

$$P_X(x) = \frac{1}{\sqrt{2\pi}} \int\limits_{-\infty}^{x} e^{-t^2/2} dt$$

$$Z = P_X\left(\frac{b-\mu}{\sigma}\right) - P_X\left(\frac{a-\mu}{\sigma}\right)$$

Then

$$h_X = -\int\limits_a^b \frac{1}{Z\sigma\sqrt{2\pi}} e^{-(x-\mu)^2/2\sigma^2} \ln\left[\frac{1}{Z\sigma\sqrt{2\pi}} e^{-(x-\mu)^2/2\sigma^2}\right] dx$$

$$= -\int_a^b \frac{1}{Z\sigma\sqrt{2\pi}} e^{-(x-\mu)^2/2\sigma^2} \left[\ln\left(\frac{1}{Z\sigma\sqrt{2\pi}}\right) - \frac{(x-\mu)^2}{2\sigma^2}\right] dx$$

$$= \ln\left(Z\sigma\sqrt{2\pi}\right) + \frac{1}{2Z\sigma^3\sqrt{2\pi}} \int_a^b (x-\mu)^2 e^{-(x-\mu)^2/2\sigma^2} dx$$

Using integration by parts with:

$$u = x - \mu \quad dv = (x - \mu)e^{-(x-\mu)^2/2\sigma^2} dx$$
$$du = dx \quad v = -\sigma^2 e^{-(x-\mu)^2/2\sigma^2}$$

yields

$$h_X = \ln\left(Z\sigma\sqrt{2\pi}\right)$$

$$+ \frac{1}{2Z\sigma^3\sqrt{2\pi}} \left[-\sigma^2(x-\mu)e^{-(x-\mu)^2/2\sigma^2}\Big|_a^b + \sigma^2 \int_a^b e^{-(x-\mu)^2/2\sigma^2} dx\right]$$

$$= \ln\left(Z\sigma\sqrt{2\pi}\right) + \frac{1}{2Z\sqrt{2\pi}}\left[\left(\frac{a-\mu}{\sigma}\right)e^{-(a-\mu)^2/2\sigma^2} - \left(\frac{b-\mu}{\sigma}\right)e^{-(b-\mu)^2/2\sigma^2}\right]$$

$$+ \frac{1}{2}\int_a^b \frac{1}{Z\sigma\sqrt{2\pi}} e^{-(x-\mu)^2/2\sigma^2} dx$$

$$= \ln\left(Z\sigma\sqrt{2\pi}\right) + \frac{1}{2Z}\left[\left(\frac{a-\mu}{\sigma}\right)\phi\left(\frac{a-\mu}{\sigma}\right) - \left(\frac{b-\mu}{\sigma}\right)\phi\left(\frac{b-\mu}{\sigma}\right)\right] + \frac{1}{2}$$

$$= \ln\left(\sigma Z\sqrt{2\pi e}e^{\frac{1}{2Z}[(\frac{a-\mu}{\sigma})\phi(\frac{a-\mu}{\sigma})-(\frac{b-\mu}{\sigma})\phi(\frac{b-\mu}{\sigma})]}\right)$$

In bits, the differential entropy becomes

$$h_X = \log_2\left(\sqrt{2\pi e}\sigma Z e^{\frac{1}{2Z}[(\frac{a-\mu}{\sigma})\phi(\frac{a-\mu}{\sigma})-(\frac{b-\mu}{\sigma})\phi(\frac{b-\mu}{\sigma})]}\right)$$

4.27　Uniform Distribution

The Uniform, or Rectangular distribution has a finite support set, and the probability density function is constant over that range, but discontinuous at the end points. This distribution appears frequently in signal processing when a sine wave signal is modeled with a random phase angle, uniformly distributed between 0 and 2π radians. An interesting property is that the sum of two independent, identically distributed uniform random variables has a symmetric triangular distribution.

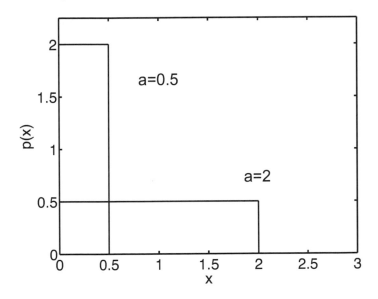

FIGURE 4.53
Probability density functions for the uniform distribution.

Probability density: $p(x) = \begin{cases} 1/a & : \quad 0 \le x \le a \\ 0 & : \quad \text{otherwise} \end{cases}$

Range: $0 \le x \le a$

Parameters: $a > 0$

Mean: $\frac{a}{2}$

Variance: $\frac{a^2}{12}$

r^{th} **moment about the origin:** $\frac{a^r}{r+1}$

Characteristic function: $(e^{iat} - 1)/iat$

Entropy: $h_X = \log_2 a$

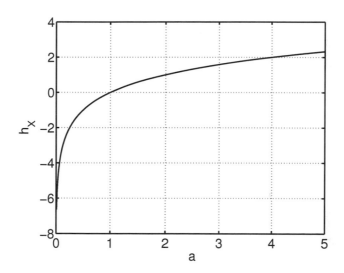

FIGURE 4.54
Differential entropy for uniform distribution.

Derivation of the differential entropy for the uniform distribution

$$h_X = -\int_0^a \frac{1}{a} \ln\left(\frac{1}{a}\right) dx$$
$$= \frac{\ln(a)}{a} \times a$$
$$= \ln(a) \quad \text{nats}$$

In bits, the entropy for the Uniform distribution becomes

$$h_X = \frac{1}{\ln(2)} (\ln(a)) = \log_2(a)$$

4.28   Weibull Distribution

The Weibull distribution is a family of two-parameter distributions with semi-infinite support, named after Waloddi Weibull although originally identified by Frechet. Special cases are the exponential distribution when $\eta = 1$ and the Rayleigh distribution when $\eta = 2$. But η need not be integral, so the Weibull distribution provides a model between these cases. For $0 < \eta < 1$ the probability density function is monotonically decreasing, whereas for $\eta > 1$ it is unimodal. If X is uniformly distributed on $[0, 1]$, then the variable $\sigma(-\ln(1 - X))^{1/\eta}$ has a Weibull distribution.

The Weibull distribution is used in reliability analyses to represent failure rates which are proportional to a power of time. It provides an appropriate model for fading channels in communications systems and for representing extreme value events such as maximum single day precipitation.

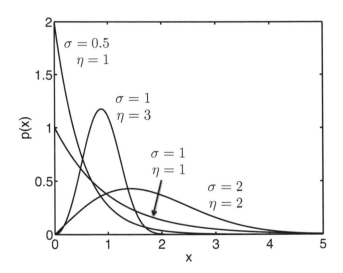

FIGURE 4.55
Probability density functions for the Weibull distribution.

Probability density: $p(x) = \begin{cases} \frac{\eta}{\sigma}\left(\frac{x}{\sigma}\right)^{\eta-1} e^{-\left(\frac{x}{\sigma}\right)^{\eta}} & : \quad 0 \leq x < \infty \\ \qquad\qquad\qquad 0 & : \quad \text{otherwise} \end{cases}$

Range: $0 \leq x < \infty$

Parameters: $\eta > 0, \sigma > 0$

Mean: $\sigma\Gamma\left(1 + \frac{1}{\eta}\right)$

Variance: $\sigma^2\left(\Gamma(1 + \frac{2}{\eta}) - [\Gamma(1 + \frac{1}{\eta})]^2\right)$

r^{th} **moment about the origin:** $\sigma^r\Gamma(1 + \frac{r}{\eta})$

Mode: $\begin{cases} 0 & : & \text{if} & \eta < 1 \\ \sigma(1 - \frac{1}{\eta})^{1/\eta} & : & \text{if} & \eta \geq 1 \end{cases}$

Entropy: $h_X = \log_2\left(\frac{\sigma}{\eta}e^{1 + \frac{\eta-1}{\eta}\gamma}\right)$

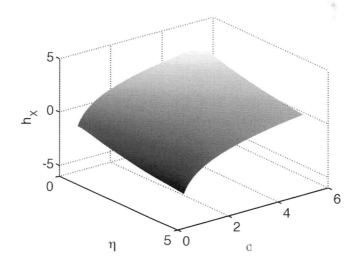

FIGURE 4.56
Differential entropy for the Weibull distribution.

Derivation of the differential entropy for the Weibull distribution

$$h_X = -\int_0^\infty \frac{\eta}{\sigma}\left(\frac{x}{\sigma}\right)^{\eta-1} e^{-\left(\frac{x}{\sigma}\right)^\eta} \ln\left[\left(\frac{\eta}{\sigma}\right)\left(\frac{x}{\sigma}\right)^{\eta-1} e^{-\left(\frac{x}{\sigma}\right)^\eta}\right] dx$$

$$= -\int_0^\infty \frac{\eta}{\sigma}\left(\frac{x}{\sigma}\right)^{\eta-1} e^{-\left(\frac{x}{\sigma}\right)^\eta} \left[\ln\left(\frac{\eta}{\sigma}\right) + (\eta-1)\ln(\frac{x}{\sigma}) - \left(\frac{x}{\sigma}\right)^\eta\right] dx$$

$$= -\ln\left(\frac{\eta}{\sigma}\right) \int_0^\infty \frac{\eta}{\sigma} \left(\frac{x}{\sigma}\right)^{\eta-1} e^{-\left(\frac{x}{\sigma}\right)^\eta} dx$$

$$- (\eta - 1) \int_0^\infty \frac{\eta}{\sigma} \left(\frac{x}{\sigma}\right)^{\eta-1} \ln\left(\frac{x}{\sigma}\right) e^{-\left(\frac{x}{\sigma}\right)^\eta} dx$$

$$+ \int_0^\infty \frac{\eta}{\sigma} \left(\frac{x}{\sigma}\right)^{2\eta-1} e^{-\left(\frac{x}{\sigma}\right)^\eta} dx$$

set $y = \frac{x}{\sigma}$ in the second and third integrals

$$= -\ln\left(\frac{\eta}{\sigma}\right) - (\eta-1) \int_0^\infty \eta y^{\eta-1} \ln(y) e^{-y^\eta} dy + \int_0^\infty \eta y^{2\eta-1} e^{-y^\eta} dy$$

set $u = y^\eta$ in both of these integrals; then $\ln(u) = \eta \ln(y)$, $du = \eta y^{\eta-1} dy$ and $u\,du = \eta y^{2\eta-1} dy$.

$$= -\ln\left(\frac{\eta}{\sigma}\right) - \frac{(\eta-1)}{\eta} \int_0^\infty e^{-u} \ln(u) du + \int_0^\infty u e^{-u} du$$

$$= \ln\left(\frac{\sigma}{\eta}\right) + \frac{(\eta-1)}{\eta} \gamma + \Gamma(2)$$

where the integral in the second term already appeared in the calculation of the entropy for the Rayleigh distribution; again γ stands for the Euler-Mascheroni constant.

$$= \ln\left(\frac{\sigma}{\eta}\right) + \frac{(\eta-1)}{\eta} \gamma + 1$$

$$= \ln\left(\frac{\sigma}{\eta} e^{1+\left(\frac{\eta-1}{\eta}\right)\gamma}\right)$$

so

$$h_X = \log_2\left(\frac{\sigma}{\eta} e^{1+\left(\frac{\eta-1}{\eta}\right)\gamma}\right) \qquad \text{bits}$$

Note that the Rayleigh distribution entropy is a special case of this result, if we set $\eta = 2$ and replace σ by $\sqrt{2}\sigma$.

5

Differential Entropy as a Function of Variance

As was stated in Chapter 2, in order to compare differential entropy for continuous probability distributions one must choose a common statistic. In this chapter the variance will serve as that statistic. This immediately excludes the Cauchy distribution since it does not have a finite variance.

In Figure 5.1 entropy is plotted as a function of variance for probability density functions which have the infinite support set $(-\infty, \infty)$. The Normal distribution has maximum entropy, as we noted in Chapter 2.

For probability density functions with semi-infinite support set $[0, \infty)$ (or $[x_o, \infty)$ in the case of the Pareto distribution) entropy as a function of variance is plotted in Figures 5.2, 5.3, and 5.4. Note that some curves are dotted lines since the generating variables take on only positive integral values and so only selective values of the variance are possible. Also some are limited in extent since the achievable values of the variance have an upper bound. The Normal curve is shown for comparison since it has maximum entropy with respect to variance for all distributions (see also [8] p. 234 and Ebrahimi et al. [10]). It is interesting to note (although we will not demonstrate it here) that if entropy were plotted as a function of the mean, the exponential distribution would maximize entropy for all distributions with semi-infinite support.

To compare entropy as a function of variance for probability density functions with finite support set, we must ensure that all support sets have the same length. In Figure 5.5 all probability density functions have support sets of unit length. When this is specified for the Uniform, Sine Wave, and Simpson distributions, the variance, and therefore the entropy, are uniquely determined, resulting in a single point on the graph. Again, the Normal distribution curve providing the maximum entropy is plotted for comparison purposes.

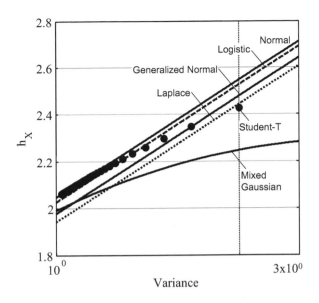

FIGURE 5.1
Differential entropy as a function of variance for probability distributions with infinite support set $(-\infty, \infty)$. Fixed parameters are $\beta = 5$ for the generalized Normal distribution and $\sigma = 0.6$ for the mixed-Gaussian distribution.

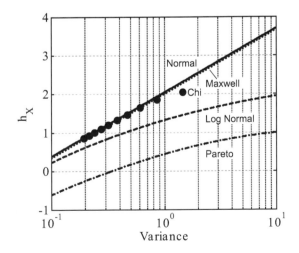

FIGURE 5.2
Differential entropy as a function of variance for the Normal distribution and some distributions with semi-infinite support set $[0, \infty)$ or $[x_o, \infty)$. Fixed parameters are $\sigma = 2$ for the Chi distribution, $m = 0.75$ for the Log Normal distribution and $x_o = 1$ for the Pareto distribution.

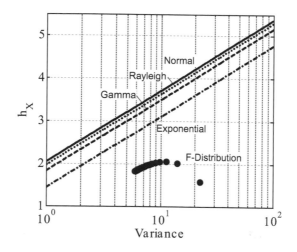

FIGURE 5.3
Differential entropy as a function of variance for the Normal distribution and some distributions with semi-infinite support set $[0, \infty)$. Fixed parameters are $\eta = 2.5$ for the Gamma distribution and $w = 5$ for the F-distribution.

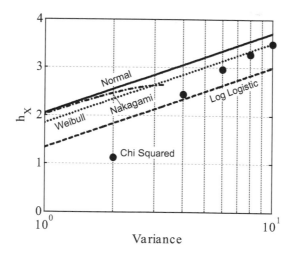

FIGURE 5.4
Differential entropy as a function of variance for the Normal distribution and some distributions with semi-infinite support set $[0, \infty)$. Fixed parameters are $\eta = 1.5$ for the Weibull distribution, $\beta = 3$ for the Log-logistic distribution and $\Omega = 9$ for the Nakagami distribution.

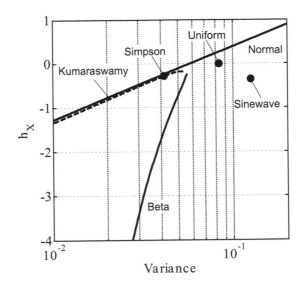

FIGURE 5.5
Differential entropy as a function of variance for the Normal distribution and
distributions with finite support set. Fixed parameters are $a = 2$ for the
Kumaraswamy distribution and $\lambda = 2$ for the Beta distribution.

6

Applications of Differential Entropy

To this point we have thoroughly discussed differential entropy and some of its properties. In this chapter we aim to provide the reader with a few applications of differential entropy to problems of interest in physics and engineering. To this point we have focused on the differential entropy of a single random variable or on the joint entropy among two random variables X, Y. In practice, however, we are typically dealing with temporal sequences of observations $x(t_1), x(t_2), \cdots, x(t_N), y(t_1), y(t_2), \cdots, y(t_N)$. From Chapter 1 we know that such sequences are modeled as random processes, that is, with the vector of random variables $\mathbf{X}$, $\mathbf{Y}$. The entropies associated with these random processes are therefore given by multi-dimensional integrals over the joint PDFs, e.g., $h_{\mathbf{X}} = \int_{\mathbb{R}^N} p_{\mathbf{X}}(\mathbf{x}) \log_2 \left(p_{\mathbf{X}}(\mathbf{x})\right) d\mathbf{x}$. Computing the entropy for a random process is therefore analytically intractable for nearly all joint distributions. An exception is the joint Gaussian distribution for which we will provide a derivation of diffential entropy shortly.

However, as we also may recall from Chapter 1, in many cases the data may be assumed stationary and each observation can usually be modeled with the same marginal PDF, hence the entropy associated with each observation will also be the same. Moreover, under the assumption of stationarity the joint PDF between any two observations becomes a one-dimensional function of only the time delay between observations. Hence, we can learn a great deal about the entropy of a random process without having to consider the full joint PDF.

Even though we will not need to treat each of the N observations separately, the fact that we have multiple observations will aid us greatly in *estimating* the differential entropy. Although this book has focused on an analytical treatment of differential entropy, some knowledge of estimation methods is necessary if we are to connect the theory to practical applications. The subject of estimation is an important one and comprises different methods for identifying our models (probabilistic or otherwise) from the observations. While a comprehensive look at estimation is not possible, the first section of this chapter aims to provide the reader with the tools needed to estimate both single and joint differential entropy.

Once the estimation methods have been established, we can turn our attention to some applications in which the differential entropy has proven useful. As we will show, differential entropy lies at the heart of several important information-theoretic quantities that have been used to great effect in signal processing and in the study of dynamical systems. Specifically we consider two such quantities: the average mutual information function, already introduced in Chapter 4, and the transfer entropy. The mutual information can be thought of as a generalization of the cross-correlation, providing a measure of common information among random processes. This measure has seen use in the detection of signals in noise [5, 11], assessing information transport in chemical and biological systems [38, 51], nonlinearity detection [39, 41], and choosing optimal sensor placement in a structural monitoring application [50]. In Section 6.2 we will demonstrate how mutual information (hence differential entropy) can be used to provide better estimates of a radar pulse delay than conventional methods.

The transfer entropy (TE) is another information-theoretic quantity that provides a slightly different definition of statistical dependency, using conditional probability to define what it means for one random process to provide information about another. This measure was originally designed to quantify information transport in dynamical systems [47] and was later extended to continuous random variables [19]. Since its introduction, the TE has been applied to a diverse set of systems, including biological [35, 47], chemical [2], economic [28], structural [39, 42], and climate [36]. In each case the TE provided information about the system that traditional linear measures of coupling (e.g., cross-correlation) could not. In Section 6.3, we explore two different applications of the transfer entropy.

6.1 Estimation of Entropy

In the preceding chapter, we have provided analytical expressions for the differential entropy of a random variable for various probability density functions. If the form of these distributions is known a priori, we may use observed data to estimate the distribution parameters. For example, consider a normally distributed random variable X, and assume we have made a sequence of independent observations $x(1), x(2), \cdots, x(N)$ of that random variable. We know from Chapter 1 that the joint distribution is

$$p_X(\mathbf{x}|\mu_X, \sigma_X) = \frac{1}{(2\pi\sigma_X^2)^{N/2}} e^{-\frac{1}{2\sigma_X^2}\sum_{n=1}^{N}(x(n)-\mu_X)^2} \tag{6.1}$$

where we have made the conditional dependence on the distribution parameters μ_X, σ_X^2 explicit. These parameters may be estimated from our observations using the familiar approach,

$$\hat{\mu}_X = \frac{1}{N} \sum_{n=1}^{N} x(n)$$

$$\hat{\sigma}_X^2 = \frac{1}{N} \sum_{n=1}^{N} (x(n) - \hat{\mu}_X)^2. \qquad (6.2)$$

It would seem that a reasonable estimate for the differential entropy might then be given by

$$\hat{H}_X = \frac{1}{2} \log_2 \left(2\pi e \hat{\sigma}_X^2 \right) \qquad (6.3)$$

that is, substitute the estimated distribution parameters into the associated entropy expression.

While this approach is intuitively appealing, we would like to be able to say something about the quality of the resulting estimates. One commonly used measure of estimator performance is the mean-square-error (MSE) between a parameter θ and the estimated value $\hat{\theta}$. This error is written mathematically

$$E[(\hat{\theta} - \theta)^2] \qquad (6.4)$$

and we would like our estimated parameter to be such that this quantity is minimized. It is easy to show ([37], page 293) that the MSE can be expanded as the sum of two terms: the estimator bias and the estimator variance. The former is given by

$$b_\theta = E[\hat{\theta}] - \theta \qquad (6.5)$$

which simply quantifies, in expectation, the degree to which our estimate deviates from the true value. We refer to an estimator as "unbiased" if $b_\theta \to 0$ and "biased" if it does not. The second component of the MSE is the estimator variance

$$var_\theta = E[(\hat{\theta} - E[\hat{\theta}])^2] \qquad (6.6)$$

which we know from Chapter 1 quantifies the spread of our estimates. An estimator that is both unbiased and possesses a small variance is therefore one that will yield values that are very close to the true value every time the estimation is performed. Returning to our estimates (6.2) we see that the bias for each is calculated as

$$b_\mu = E\left[\frac{1}{N}\sum_{i=1}^{N} x(i)\right] - \mu_X = \frac{1}{N}\sum_{i=1}^{N} E\left[x(i)\right] - \mu_X = \frac{N\mu_X}{N} - \mu_X = 0$$

$$b_{\sigma^2} = E\left[\frac{1}{N}\sum_{i=1}^{N}\left((x(i) - \mu_X) - \frac{1}{N}\sum_{j=1}^{N}(x(j) - \mu_X)\right)^2\right] - \sigma_X^2$$

$$= \frac{1}{N}\sum_{i=1}^{N} E\left[(x(i) - \mu_X)^2\right] - \frac{2}{N^2}\sum_{i=1}^{N} E\left[(x(i) - \mu_X)^2\right]$$

$$+ \frac{1}{N^3}\sum_{i=1}^{N}\sum_{j=1}^{N} E\left[(x(j) - \mu_X)^2\right] - \sigma_X^2$$

$$= -2\frac{1}{N}\sigma_X^2 + \frac{1}{N}\sigma_X^2 = -\frac{\sigma_X^2}{N}. \tag{6.7}$$

Thus, while our mean estimate is unbiased, our variance parameter estimate is not (had we normalized 6.2 by $N-1$ instead of N the estimator *would* have been unbiased). However, it is *asymptotically* unbiased, that is to say,

$$\lim_{N\to\infty} E[\hat{\sigma}^2] - \sigma^2 = 0 \tag{6.8}$$

so that for large datasets we can expect good results. The variance of each estimator (6.6) can be similarly calculated and is found to be [29]

$$var_\mu = \frac{\sigma^2}{N}$$

$$var_{\sigma^2} = \frac{2\sigma^4}{N}. \tag{6.9}$$

Again we see that asymptotically (for large N) the estimators will perform well, possessing a vanishingly small variance. An estimator for which this property holds is referred to as *consistent*. Although our estimator is consistent, other estimators may exist with a smaller variance than that given by (6.9). In fact, given the joint distribution of the data, $p_X(\mathbf{x}|\boldsymbol{\theta})$ (e.g., 6.1) we can derive a lower bound on the minimum possible error variance for any estimator possessing either a constant or zero bias (see e.g., [20 and 29]). Given an unknown parameter vector $\boldsymbol{\theta} = (\theta_1, \theta_2, \cdots, \theta_M)$, define the $M \times M$ Fisher information matrix

$$I_{ij}(\boldsymbol{\theta}) = -E\left[\frac{\partial^2 \ln\left(p_X(\mathbf{x}|\boldsymbol{\theta})\right)}{\partial\theta_i\partial\theta_j}\right]. \tag{6.10}$$

Then

$$E[(\hat{\theta}_i - \theta_i)^2] \geq I^{-1}(\boldsymbol{\theta})_{ii} \tag{6.11}$$

The limit implied by Eqn. (6.11) is referred to as the "Cramér-Rao Lower Bound" (CRLB) and is the minimum conditional error variance possible for an un-biased estimator (note the CRLB for a biased estimator can also be calculated and is given on page 400 of [29]). In keeping with our joint normal model (6.1) we have

$$
I(\boldsymbol{\theta}) = -E \left[\begin{array}{cc} -\frac{N}{\sigma^2} & -\frac{1}{\sigma^4}\sum_{i=1}^{N} x(i) - \mu_X \\ -\frac{1}{\sigma^4}\sum_{i=1}^{N} x(i) - \mu_X & \frac{N}{2\sigma^4} - \frac{1}{\sigma^6}\sum_{i=1}^{N}(x(i) - \mu_X)^2 \end{array} \right]
$$
$$
= \left[\begin{array}{cc} \frac{N}{\sigma^2} & 0 \\ 0 & \frac{N}{2\sigma^4} \end{array} \right] \tag{6.12}
$$

so that upon inversion, we have the CRLB for both parameters

$$
E[(\hat{\mu}_X - \mu_X)^2] \geq \frac{\sigma^2}{N}
$$
$$
E[(\hat{\sigma}_X - \sigma_X)^2] \geq \frac{2\sigma^4}{N}. \tag{6.13}
$$

This demonstrates that our estimators (6.2) achieve the CRLB and are therefore referred to as *efficient* estimators. In the sense of minimizing (6.4) an estimator that is both unbiased and efficient is the best we can hope to do.

At this point we have described the basic properties needed to evaluate the quality of our parameter estimates. Now, returning to the problem of entropy estimation we would like to test the proposed estimator (6.3) using our asymptotically unbiased and efficient estimator for σ_X^2, i.e., Eqn. (6.2). Consider a sequence of N independent, Gaussian distributed random variables with zero mean and $\sigma_X^2 = 1$. Substituting (6.2) in (6.3) we plot the resulting $\hat{h}_X$ as a function of the number of samples in Figure 6.1. As predicted, the quality of the estimate improves with increasing data. Moreover the estimate converges to the correct value (is unbiased) and shows minimal variance about the true value for large N.

It turns out that the entropy estimation procedure we have just described is both asymptotically unbiased and efficient. The reason for this is that in estimating the parameters (6.2) we have used the method of Maximum Likelihood. Maximum Likelihood estimation is a powerful approach to identifying model parameters from observed data. Thorough treatment of this approach is provided in [37]. In short, the approach takes the parameter estimates to be those that maximize the probability of having observed the sequence $x(1), x(2), \cdots, x(N)$, i.e., maximize (6.1). Noting that the logarithm is a monotonically increasing function, we could equivalently maximize the log of (6.1). Taking the derivative of (6.1) with respect to μ_X, σ_X^2, setting each equal to zero and solving for the unknown parameters yields precisely the estimators given by (6.2).

The power in using MLEs is that they are guaranteed to be both asymptotically unbiased and efficient (see [20], Theorem 7.3, page 183). Moreover, there

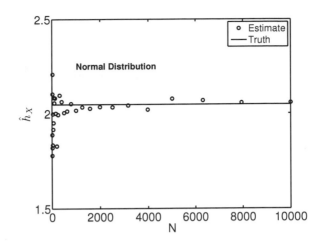

FIGURE 6.1
Estimated entropy for the Normal distribution as a function of the number of samples used in the estimate, N. The true value is given by the solid line.

is an "invariance" theorem (see Ref. [38], page 284) that states that a function of quantities for which we have an MLE will also be an MLE. That is to say, if $\hat{\theta}$ is an MLE, an MLE of $g(\theta)$ is simply $g(\hat{\theta})$ [37]. Thus, our approach of substituting the MLEs for the distribution parameters into (6.3) will produce an MLE for the differential entropy. Our approach is therefore on firm theoretical footing and will produce the best possible estimate (in the sense of minimizing 6.4), provided that we have collected enough samples. This approach can be used in conjunction with any of the differential entropies derived herein.

As a second example we consider both the Laplace and Rayleigh distributions with true parameter values $\lambda = 3$ (Laplace) and $\sigma = 2$ (Rayleigh), respectively. The differential entropies associated with these distributions were given in Chapter 4 as $h_X = \log_2(2e/\lambda) = 0.858$ and $h_X = \log_2(\frac{\sigma}{\sqrt{2}}e^{1+\gamma/2}) = 2.359$. For each distribution, the MLEs for the parameters λ, σ are straightforward to derive. We simply maximize the PDF of each distribution with respect to the distribution parameter. The MLE for the Laplace shape parameter is found to be

$$\hat{\lambda} = \frac{1}{N} \sum_{n=1}^{N} |x(n)| \tag{6.14}$$

which can be substituted into the expression for the differential entropy giving $\hat{h}_X = \log_2(2e/\hat{\lambda})$. The MLE of the shape parameter for the Rayleigh distribu-

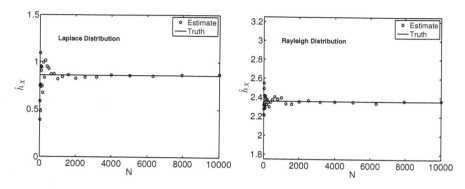

FIGURE 6.2
Estimated differential entropy for the Laplace (left) and Rayleigh (right) distributions as a function of the number of samples used in the estimate, N. The true value is given by the solid line.

tion can also be derived as

$$\hat{\sigma} = \sqrt{\frac{1}{2N} \sum_{n=1}^{N} x^2(n)} \tag{6.15}$$

giving the estimated entropy $\hat{h}_X = \log_2(\frac{\hat{\sigma}}{\sqrt{2}} e^{1+\gamma/2})$. Figure 6.2 shows the improvement in the estimates as a function of the number of data N. Again, we see convergence to the true value as N is increased. Finally, we should point out that nothing in what we have just described restricts us to continuous probability distributions; for example, a random process where each observed value is chosen independently from the discrete Poisson distribution (3.1). In this case the likelihood is

$$\ln\left(\frac{\lambda^{\sum_{n=1}^{N} x(n)} e^{-N\lambda}}{(x(n)!)^N}\right)$$

$$= \sum_{n=1}^{N} x(n) \ln(\lambda) - N\lambda - \ln\left[\prod_{n=1}^{N} x(n)!\right] \tag{6.16}$$

so that differentiating with respect to λ and setting equal to zero yields the MLE

$$\hat{\lambda} = \frac{1}{N} \sum_{n=1}^{N} x(n). \tag{6.17}$$

Now assume a large value of $\lambda = 7$ and use the approximation [12] to estimate the entropy (to order $1/\lambda^2$) as $\hat{H}_X \approx \frac{1}{2}\ln(2\pi e\hat{\lambda}) - 1/\hat{\lambda} - 1/(24\hat{\lambda}^2) - 19/(360\hat{\lambda}^3)$. Figure 6.3 shows the estimate for the Poisson entropy as a function of the

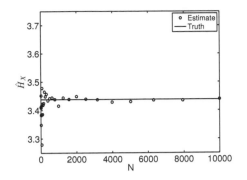

FIGURE 6.3
Estimated differential entropy for the Poisson distribution as a function of
the number of samples used in the estimate, N. The true value (to good
approximation) is given by the solid line.

number of data. Whether we are interested in discrete or continuous probability models, the MLE approach to entropy estimation can nearly always be
expected to yield good results.

The problem with the above-described approach is that in some instances
we do not know the underlying probability density and instead have only
the observations. Additionally, our parametric approach is not of much use
for joint entropies as we do not have a closed-form solution for those, the
sole exception being jointly Gaussian distributed random variables. Thus, we
are often left with the task of estimating single and joint entropies from observed data without recourse to a known probability model. This problem is
addressed next.

Given our sequence of observations, assumed to originate from a stationary
random process, we would like to estimate

$$h_X = -\int_X p_X(x) \log_2 (p_X(x))\, dx$$

$$h_{XY} = -\int_Y \int_X p_{XY}(x,y) \log_2 (p_{XY}(x,y))\, dx dy. \qquad (6.18)$$

We have already noted in Section 2.1 the differential entropies are equivalently
written as expected values w.r.t. the "$\log(\cdot)$" function, that is to say

$$h_X = -E[\log_2 (p_X(x))]$$

$$h_{XY} = -E[\log_2 (p_{XY}(x))]. \qquad (6.19)$$

There exists a straightforward means of approximating expectations given a
stationary sequence of observed data $x(n)$ $n = 1 \cdots N$. The ergodic Theorem
of Birkhoff [6] states that for large N we may replace the expectation operator

by a summation over the observations. More specifically,

$$\lim_{N \to \infty} \frac{1}{N} \sum_{n=1}^{N} g(x(n)) = \int_{X} g(x(n)) p_X(x(n)) dx = E[g(x(n))]. \qquad (6.20)$$

The function $g(\cdot)$ is a deterministic function of the random variable X, in our case, $g(x) = \log_2(p_X(x))$. Thus, our estimators become

$$\hat{h}_X = -\frac{1}{N} \sum_{n=1}^{N} \log_2\left(\hat{p}_X(x(n))\right)$$

$$\hat{h}_{XY} = -\frac{1}{N^2} \sum_{n=1}^{N} \sum_{m=1}^{N} \log_2\left(\hat{p}_{XY}(x(n), y(m))\right) \qquad (6.21)$$

which are nothing more than simple averages over the available data. However, we still require a means of estimating $p_X(x(n))$ and $p_{XY}(x(n), y(m))$, i.e., the single and joint probability density functions. Perhaps the most straightforward estimator is to simply divide X into B discrete "bins," where the start of each bin is given by $x(i) = \min(\mathbf{x}) + (i-1)\Delta_x$, $i = 1 \cdots B$. The bin width $\Delta_x = (\max(\mathbf{x}) - \min(\mathbf{x}))/(B-1)$ is a parameter to be chosen by the practitioner (alternatively we might specify the number of bins and determine Δ_x). Now denote the number of data falling in the ith bin as $b_i = \#(x(i) \le x(n) < x(i+1))$, $n \in [1, N]$. As the number of data becomes large, one would expect

$$\hat{p}_X(x(i)) = \frac{b_i}{N\Delta_x} \qquad (6.22)$$

to be a reasonable estimator of the probability density at the location defined by the bin, $x(i)$. This estimator is further assumed uniformly valid for any point $x(n)$ lying in the ith bin. Note that this estimator closely resembles the definition of a PDF (1.5), that is, the probability of finding an observation in a vanishingly small interval Δ_x. Similarly for the joint PDF one chooses a bin width Δ_y and counts b_{ij} as $\#(x(i) \le x(n) < x(i+1)$ and $y(j) \le y(m) < y(j+1))$ yielding the estimator

$$\hat{p}_{XY}(x_i, y_j) = \frac{b_{ij}}{N\Delta_x\Delta_y} \qquad (6.23)$$

The estimators (6.22) and (6.23) are the familiar histogram estimators. The validity of these estimators is obviously predicated on the assumption of stationarity so that the marginal distribution associated with any particular observation is the same, regardless of where in the signal it occurs. It is also clearly dependent on the size of the dataset N and the choice of bin width. A great deal of effort has been focused on density estimation; a good reference on the topic is provided by Silverman [49]. For the histogram estimator it is suggested that the number of bins be chosen $B \approx \sqrt{N}$.

Given these estimators for the PDF we have finally, for the single and joint entropies

$$\hat{h}_X = -\frac{1}{B} \sum_{i=1}^{B} \hat{p}_X(x(i))$$

$$\hat{h}_{XY} = -\frac{1}{B_x B_y} \sum_{i=1}^{B_x} \sum_{j=1}^{B_y} \hat{p}_{XY}(x(i), y(j)). \tag{6.24}$$

Because the value of the PDF estimate is assumed uniform for all points in the ith bin the summation in Eqn. (6.21) is over all B bins rather than all N points.

This is perhaps the simplest entropy estimator available but is nonetheless effective as will be shown in the next section. However, better estimators are available and rely, not surprisingly, on improved estimators of the PDFs. Rather than use a fixed grid of bins over the space of observations, we may allow each data value to play the role of a bin center. That is to say, given that we would like an estimate of $p_X(x(n))$ it makes sense to look at the frequency with which data populate a local neighborhood about $x(n)$. More specifically, define the *kernel density* estimator for the PDF as

$$\hat{p}_X(x(n)) = \frac{1}{N \Delta_x} \sum_{i=1}^{N} K \left(\frac{x(i) - x(n)}{\Delta_x} \right) \tag{6.25}$$

where $K \left(\frac{x(i)-x(n)}{\Delta_x} \right)$ is a kernel function that more heavily weights the contribution of data near $x(n)$. This is a generalization of the histogram estimator which weighted all samples within Δ_x of $x(i)$ uniformly. If many data are near $x(n)$ we will estimate a higher probability near this value and vice versa. Again, this corresponds to our intuition about probability. One commonly used kernel is the Heaviside kernel

$$K_\Theta \left(\frac{x(i) - x(n)}{\Delta_x} \right) = \Theta \left(\frac{1}{2} - \frac{|x(i) - x(n)|}{\Delta_x} \right) \tag{6.26}$$

where

$$\Theta(x) = \left\{ \begin{array}{ll} 1 & : \quad x \geq 0 \\ 0 & : \quad x < 0 \end{array} \right. \tag{6.27}$$

which simply counts all points within $\Delta_x/2$ of the point $x(n)$. The number of "bins" using this approach is the number of data N as each point is considered in turn. Other popular kernel choices include Gaussian and triangular shaped functions. Regardless of the particular functional form, the practitioner still is required to set the kernel size Δ_x, sometimes referred to as the "kernel bandwidth" in the literature [49]. The extension to multiple dimensions is straightforward. In two dimensions, for example, we have the estimator

$$\hat{p}_{XY}(x(n), y(m)) = \frac{1}{N\Delta_x\Delta_y} \sum_{i=1}^{N} \sum_{j=1}^{N} K\left(\sqrt{\left(\frac{x(i) - x(n)}{\Delta_x}\right)^2 + \left(\frac{y(j) - y(m)}{\Delta_y}\right)^2}\right)$$

(6.28)

which uses the norm of the distance vector in the argument of the kernel. If using the Heaviside kernel, for example, this approach simply counts the number of points within a radius $\sqrt{\Delta_x^2 + \Delta_y^2}$ of the point $x(n)$.

Yet another interesting twist on the kernel-based approach is to consider an adjustable bandwidth. Let $\Delta_x(x(n))$ be a function of the data and instead fix a small number of points $M \ll N$. The idea is to find the minimum volume element required to encompass the nearest M points. More specifically, define the one-dimensional volume element

$$\Delta_x(x(n)) = \Delta_x : \sum_{i=1}^{N} K_{\Theta}\left(\frac{x(i) - x(n)}{\Delta_x}\right) = M \qquad (6.29)$$

The PDF estimate at $x(n)$ is then given by

$$\hat{p}(x(n)) = \frac{1}{N}\frac{M}{\Delta_x(x(n))} \qquad (6.30)$$

and is referred to as a "fixed mass" kernel. As with the fixed-bandwidth approach, the extension to higher dimensions is straightforward. In the two-dimensional case one finds the minimum volume element, centered at $(x(n), y(n))$, containing M points. The estimate is then given by

$$\hat{p}(x(n), y(m)) = \frac{1}{N}\frac{M}{A(x(n), y(m))} \qquad (6.31)$$

where

$$A(x(n), x(m)) = \Delta_x\Delta_y : \sum_{i=1}^{N}\sum_{j=1}^{N} K_{\Theta}\left(\frac{x(i) - x(n)}{\Delta_x}\right) K_{\Theta}\left(\frac{x(j) - y(m)}{\Delta_y}\right) = M$$

(6.32)

is the smallest area element in which M points can be found. Higher dimensional volume elements are defined accordingly.

Whereas our parametric estimation approach is on solid theoretical grounds, the above-described non-parameteric approaches are not. We would like to be able to declare our density estimates to be MLEs in which case, by the aforementioned invariance property, so too would be our entropy estimates (at least for large N). Only a scale parameter (Δ_x or M) typically appears in the kernel-based approaches and we have given no guidance on selection.

It turns out that the best we can do is to keep these parameters "as small as possible." This owes to the fact that as the number of data increase, and the scale parameters shrink, the density estimates (hence the entropy estimates) converge to their true value [19]. The key question is therefore how far can one shrink the kernel bandwidth and still retain enough points in the "bins" to yield a good estimate. This question is unanswered as far as we are aware; however, we will present results to suggest that high quality, non-parametric estimates of differential entropy are indeed possible.

6.2   Mutual Information

As we have already shown in Chapter 2, the mutual information function quantifies common information between two random variables. To review, recall from Chapter 1 that two random variables are said to be independent if their joint probability distribution factors as $p_{XY}(x,y) = p_X(x)p_Y(y)$. One way to quantify deviations from this assumption is to compute the scalar [23]

$$I_{XY} = \int_X \int_Y p_{XY}(x,y) \log_2 \left(\frac{p_{XY}(x,y)}{p_X(x)p_Y(y)} \right) dx dy$$
$$= h_X + h_Y - h_{XY} \tag{6.33}$$

By inspection, we see that if the random variables are indeed independent we have $I_{XY} = 0$ rising to a maximum value if one random variable is a deterministic function of the other. The mutual information function is strictly non-negative and it is assumed that $0 \log_2[0] = 0$.

6.2.1   Mutual Information for Random Processes

Our earlier discussions of the mutual information took place in the context of two random variables; however, in many practical applications we are interested in the properties of two different random *processes*, **X**, **Y** (see Section 1.3) which are used to model the sequences of observations $\mathbf{x} = (x(t_1), x(t_2), \cdots, x(t_N))$ and $\mathbf{y} = (y(t_1), y(t_2), \cdots, y(t_N))$, respectively. Here, as is often the case, the random variables are indexed by sampling times $t_1, t_2, \cdots, t_N$, each assumed to be separated by the constant sampling interval Δ_t. Given such a collection, it is conceivable that we might want to compute the mutual information between these two collections.

    One approach would be to use the estimation methods described in Section 6.1. Indeed, we will make use of the basic histogram estimator shortly; however, it is first instructive to consider the case where an expression may be obtained analytically. The only multivariate probability distribution that readily admits an analytical solution for the differential entropies is the jointly

Gaussian distribution. The Gaussian models for both the individual and joint data vectors are

$$p_{\mathbf{X}}(\mathbf{x}) = \frac{1}{(2\pi)^{N/2}|C_{\mathbf{XX}}|^{1/2}} e^{-\frac{1}{2}(\mathbf{x}-\mu_{\mathbf{x}})^T C_{\mathbf{XX}}^{-1}(\mathbf{x}-\mu_{\mathbf{x}})}$$

$$p_{\mathbf{Y}}(\mathbf{y}) = \frac{1}{(2\pi)^{N/2}|C_{\mathbf{YY}}|^{1/2}} e^{-\frac{1}{2}(\mathbf{y}-\mu_{\mathbf{y}})^T C_{\mathbf{YY}}^{-1}(\mathbf{y}-\mu_{\mathbf{y}})}$$

$$p_{\mathbf{XY}}(\mathbf{x},\mathbf{y}) = \frac{1}{(2\pi)^N|C_{\mathbf{XY}}|^{1/2}} e^{-\frac{1}{2}(\mathbf{x}-\mu_{\mathbf{x}})^T C_{\mathbf{XY}}^{-1}(\mathbf{y}-\mu_{\mathbf{y}})} \tag{6.34}$$

where $C_{\mathbf{XX}}$, $C_{\mathbf{YY}}$ are the $N \times N$ covariance matrices associated with the random processes $\mathbf{X}$, $\mathbf{Y}$, respectively and $|\cdot|$ takes the determinant. The matrix $C_{\mathbf{XY}}$ is the $2N \times 2N$ covariance matrix associated with the joint data vector $[\mathbf{x}, \mathbf{y}]$.

If the joint distribution is given by Eqn. (6.34) the entropy for the random process $\mathbf{X}$ (for example) is

$$h_{\mathbf{X}} = -E[\log(p_{\mathbf{X}}(\mathbf{x}))]$$

$$= E[\log\left(2\pi^{N/2}|\mathbf{C}_{\mathbf{XX}}|^{1/2}\right)] + E[\frac{1}{2}(\mathbf{x}-\mu_{\mathbf{X}})^T \mathbf{C}_{\mathbf{XX}}^{-1}(\mathbf{x}-\mu_{\mathbf{X}})]$$

$$= \frac{1}{2}\log_2((2\pi)^N) + \frac{1}{2}|\mathbf{C}_{\mathbf{XX}}| + \frac{1}{2}E[(\mathbf{x}-\mu_{\mathbf{X}})^T \mathbf{C}_{\mathbf{XX}}^{-1}(\mathbf{x}-\mu_{\mathbf{X}})]$$

$$= \frac{1}{2}\log_2((2\pi)^N) + \frac{1}{2}|\mathbf{C}_{\mathbf{XX}}| + \frac{N}{2}$$

$$= \frac{1}{2}\log\left((2\pi e)^N|\mathbf{C}_{\mathbf{XX}}|\right) \tag{6.35}$$

and similarly for $\mathbf{Y}$

$$h_{\mathbf{Y}} = \frac{1}{2}\log\left((2\pi e)^N|\mathbf{C}_{\mathbf{YY}}|\right). \tag{6.36}$$

The joint entropy is likewise governed by $C_{\mathbf{XY}}$ and is given by $h_{\mathbf{XY}} = \frac{1}{2}\log\left((2\pi e)^{2N}|\mathbf{C}_{\mathbf{XY}}|\right)$, thus we have for the mutual information

$$I_{\mathbf{XY}} = \frac{1}{2}\log\left(\frac{|\mathbf{C}_{\mathbf{XX}}||\mathbf{C}_{\mathbf{YY}}|}{|\mathbf{C}_{\mathbf{XY}}|}\right). \tag{6.37}$$

While this special case is certainly of a simpler form (in particular we have eliminated the integrals in 6.33) it still requires the determinants of potentially high-dimensional (large N) matrices. Now, in many signal processing applications the random processes are assumed to be stationary where each observation is modeled with the same marginal PDF. As we have seen in Chapter 1, for a stationary random process each of the entries in $\mathbf{C}_{\mathbf{XY}}$ can be indexed by a single discrete time delay, i.e., $C_{XY}(\tau) \equiv E[x(t_n)y(t_n+\tau)]$. What this means is that the mutual information between any two samples from X and Y depends *only* on the temporal separation τ. Thus, for stationary random processes much can be learned by simply focusing on the mutual information between any two given observations from X and Y. In this case (6.37) becomes

$$I_{XY}(\tau) = \frac{1}{2} \log \left(\frac{\sigma_X^2 \sigma_Y^2}{\sigma_X^2 \sigma_Y^2 - E[x(t_n)y(t_n + \tau)]} \right)$$

$$= -\frac{1}{2} \log \left(1 - \rho_{XY}^2(\tau) \right) \tag{6.38}$$

which will be referred to as the time-delayed mutual information function for stationary, jointly Gaussian distributed signals. Note this is just a simple extension of the mutual information associated with jointly Gaussian random variables (see Ex. 7 of Chapter 2, Eqn. 2.20). In this special case, the mutual information is a function of the linear cross-correlation coefficient, $\rho_{XY}(\tau)$.

Returning to the general case (arbitrary PDFs), the mutual information between any two observations taken from stationary random processes X and Y can be written (using the law of conditional probability 1.20)

$$I_{XY}(\tau) = \int \int p_{XY}(x(t_n), y(t_n + \tau)) \log_2 \left(p_{XY}(x(t_n), y(t_n + \tau)) \right) dx dy$$

$$- \int p_X(x(t_n)) \log_2 \left(p_X(x(t_n)) \right) dx$$

$$- \int p_Y(y(t_n)) \log_2 \left(p_Y(y(t_n)) \right) dy$$

$$= h_X + h_Y - h_{XY}(\tau) \tag{6.39}$$

which, due to the assumption of stationarity, is a function of τ only. We have also implicitly noted that stationarity allows us to write $p_Y(y(t_n)) = p_Y(y(t_n + \tau))$ so that only the joint entropy is a function of delay. In light of (6.38), the time-delayed mutual information can be thought of as a general measure of correlation between two random processes. It is a function of both the single and joint differential entropies associated with the random variables X and Y.

6.2.2   Radar Pulse Detection

Given this background, we consider the problem of detecting the time-of-flight of a radar pulse. In this application a known (clean) pulse is sent from a transmitter, is reflected by an object of interest, and received at the sender's location. Precise knowledge of the time to return (given the known propagation speed of the pulse) is sufficient to determine the range to the target. However, in transit the pulse becomes corrupted by various sources of "clutter," thus the estimation of the delay time is a challenging problem.

Denote the pristine (known) waveform $x(t)$ and the return signal $y(t)$. Further assume that the clutter is additive in nature such that we have the model

$$y(t) = x(t - \tau) + \eta(t) \tag{6.40}$$

where $\eta(t)$ is a stationary random process describing the clutter at time t with PDF $p_H(\eta)$. As we have done throughout, we assume a sampled version of (6.40) of the received waveform $y(t_n) = x(t_n - \tau) + \eta(t_n)$ $n = 1 \cdots N$ with sampling interval Δ_t. If the source of the clutter is modeled as a joint normally distributed random process, it can be shown that the MLE for time delay is

$$\hat{\tau} = \max_{\tau} \; \hat{\rho}_{XY}(\tau) \tag{6.41}$$

i.e., the maximum of the time-delayed cross-correlation coefficient, Eqn. (1.56) of Chapter 1. However, if the noise is *not* jointly Gaussian distributed, claims of optimality cannot be made using this approach.

It has long been recognized, however, that the time-delayed mutual information provides a more general alternative that can be used effectively in such situations (see, e.g., [11, 27]). In fact, in [45], it was shown that maximizing the time-delayed mutual information function provides an MLE, *regardless* of the probability distribution of the noise source $(p_H(\eta))$, so long as the noise samples are independent, identically distributed. Thus, the alternative time-delay estimator becomes

$$\hat{\tau} = \max_{\tau} \; \hat{I}_{XY}(\tau). \tag{6.42}$$

In the special case of jointly Gaussian observations we see from Eqn. (6.38) that this function is maximized by the same value of τ that maximizes $\rho_{XY}(\tau)$. However, as will be shown, the maximum MI approach yields considerably better estimates, particularly for strongly non-Gaussian distributions. The reader may also notice that the maximizer of (6.42) is really just the minimizer of the joint differential entropy (see Eqn. 6.39); hence, an equivalent estimator would be $\hat{\tau} = \min_{\tau} \hat{h}_{XY}(\tau)$.

To illustrate, consider the pulse shown in Figure 6.4. The pulse is given by the function

$$x_{\tau}(t) = Ae^{-(t-\tau)^2/(2\sigma^2)} \sin(2\pi f(t - \tau) + \phi) \tag{6.43}$$

where, in this example, the frequency of the pulse is $f = 5 \; GHz$ and the pulse width $\sigma = 1.58e^{-9}$. In what follows the variance of the noise signal will remain fixed at $\sigma_H^2 = 1$. Thus, the amplitude of the pulse will be dictated by the signal-to-noise ratio (SNR) which can be shown to be $A = 3.92 \times 10^{SNR/20}$ [45]. The lower the SNR, the lower the amplitude.

We consider each noise value, i.e., each $\eta(t_n)$, to be independent and identically distributed and consider two governing distributions. The first distribution considered is the generalized normal distribution already covered in Chapter 4. Recognizing that $\eta(t_n) = y(t_n) - x(t_n)$, the PDF for the sequence of observations gives the likelihood

$$p_Y(\mathbf{y}|\tau) = \prod_{n=0}^{N-1} \frac{\beta}{2\alpha\Gamma\left(\frac{1}{\beta}\right)} e^{-(|y(t_n)-x_{\tau}(t_n)|/\alpha)^{\beta}} \tag{6.44}$$

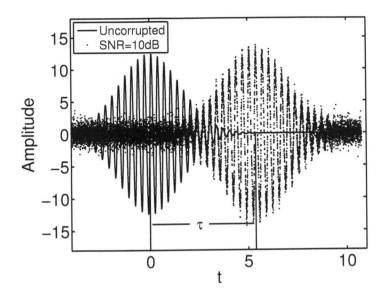

FIGURE 6.4
Clean and corrupted (return) pulse with an SNR of 10dB. The goal of the estimator is to find the time τ between the time the pulse is sent and the time at which it returns.

for parameters σ, $\beta > 0$. As the shape parameter $\beta \to \infty$, the generalized normal distribution approaches the Uniform distribution (see Generalized Normal Distribution in Chapter 4). However, unlike the Uniform distribution, the generalized normal possesses continuous support $[-\infty, \infty]$. This is important as it allows one to derive the CRLB on the variance of the error of an unbiased estimator, $E[(\hat{\tau} - \tau)^2]$. As we have already described in Section 6.1, the CRLB holds an important place in estimation theory as the minimum obtainable mean square error for an unbiased estimator. Although it does not guarantee the existance of such an estimator, it represents a fundamental limit on the best one can hope to do in a given estimation problem.

The Fisher information associated with an estimate for τ is found to be [45]

$$I_{GN}(\tau) = \frac{\left(\alpha^{2\beta} + (-\alpha)^{2\beta}\right) \Gamma\left(2 - \frac{1}{\beta}\right)}{2\alpha^2 \Gamma\left(\frac{1}{\beta}\right)} \sum_{n=0}^{N-1} \left(\frac{dx(t_n)}{d\tau}\right)^2 \tag{6.45}$$

and we have that $E[(\hat{\tau} - \tau)^2] \geq 1/I_{GN}(\tau)$, assuming our estimator is unbiased. The requirement that the noise be of unit variance mandates that we set

$$\alpha = \sqrt{\frac{\Gamma(1/\beta)}{\Gamma(3/\beta)}}. \tag{6.46}$$

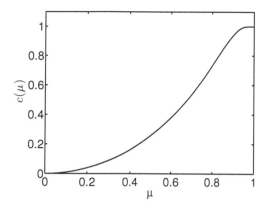

FIGURE 6.5
Function $c(\mu)$ in the Mixed-Gaussian FI plotted as a function of the shape parameter μ.

As a second corrupting noise source we choose the mixed-Gaussian distribution (see Mixed-Gaussian Distribution in Chapter 4). This distribution also admits analytical calculation of the Fisher information

$$I_{MG}(\tau) = \frac{\sigma^2 + \mu^2(c(\mu) - 1)}{(\sigma^2 + \mu^2)^2} \sum_{n=0}^{N-1} \frac{dx_\tau(t_n)}{d\tau}^2$$

$$= \frac{1 + (c(\mu) - 2)\mu^2}{(1 - \mu^2)^2} \sum_{n=0}^{N-1} \frac{ds_\tau(t_n)}{d\tau}^2 \tag{6.47}$$

where in the last step we have substituted $\sigma^2 = 1 - \mu^2$ for the scale parameter to enforce unit variance on the noise vector. The function $c(\mu)$ is the result of numerically taking the expectation (integrating) $\left\{\tanh\left[\frac{\mu\eta(t_n)}{1-\mu^2}\right]^2\right\}$ with respect to the mixed-Gaussian density. This function, shown in Figure 6.5 depends on μ only and determines the degree to which the Fisher information for the mixed-Gaussian distribution deviates from that associated with the standard Gaussian noise model. The Fisher information for the Gaussian case is recovered for $\mu = 0$ as expected while for $\mu \to 1$ the CRLB$\to 0$. As the peaks separate ($\mu \to 1$) the variance of the individual peaks must tend to zero, again due to the constraint that $E[\eta(t_n)\eta(t_n)] = 1$. In this limit the noise PDF becomes two delta functions located at $\mu = \pm 1$, thus the uncertainty in the observations tends to zero as does the CRLB.

We now have the CRLB as a function of the signal-to-noise ratio and can therefore assess the quality of the maximum mutual information estimator (6.42) vs. the standard estimator (6.41) for two different non-Gaussian noise models. As a numerical experiment, we generate signals according to the following real-valued sinusoidal "pulse" (6.43). In all experiments, the signals consist of 2^{14} points sampled at $\Delta_t = 0.0013e^{-9}$ while the frequency and

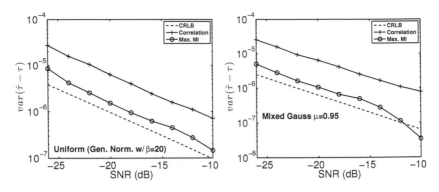

FIGURE 6.6
Mean square error in the pulse delay estimate as a function of SNR for uniform distributed noise (left) and Mixed-Gaussian noise with $\mu = 0.95$ (right).

scale parameters were taken to be $f = 5e^{-9}$, $\alpha = 1.58e^{-9}$, respectively. The numerical experiment proceeds by selecting τ from a uniform distribution in the interval of $[0, 2^{14} - 1]\triangle_t$ (we use zero padding when samples out of this range are needed for computation) and computing $x(t_n)$ as in Eqn. (6.40), for the corresponding noise model. The SNR is controlled entirely by the signal amplitude A (recall the noise variance is always unity).

In the case of the generalized normal we use $\mu = 0$, $\alpha = 1.77$, $\beta = 20$ which produces a nearly uniform distribution with unit variance. For the mixed-Gaussian case we choose $\mu = 0.95$, $\sigma^2 = 0.098$ to produce a two-peaked distribution, again with unit variance. We then generated clean pulses of varying amplitude so that a variety of SNR values were achieved. For each we used the maximum MI estimator (6.42) where the basic histogram approach was used to provide $\hat{I}_{X,Y}(\tau)$ for each possible delay $\tau_n = n\Delta_t$. For the histogram we used $B = \sqrt{N} = 128$ bins in the estimate. Results for this noise case are shown in Figure 6.6. Again, the CRLB curves represent the best theoretical performance one can obtain for this problem. Given that our estimator attempts to maximize the empirical likelihood, it is not surprising that we come close to attaining the CRLB. The Maximum Mutual Information estimator clearly outperforms the cross-correlation approach across all values of SNR. The performance gains are non-negligible, showing nearly an order of magnitude increase in performance for certain SNR values. We should point out that there is a downward bias in the estimator when the standard deviation of the error approaches the size of the sampling interval [45]. This bias is, in fact, the reason the curve appears to dip below the CRLB which assumed an unbiased estimator. Although the estimation of the mutual information is certainly more challenging than estimating the cross-correlation coefficient, this study clearly demonstrates the power of differential entropy.

6.3 Transfer Entropy

The time-delayed mutual information has been used in several applications as a measure of information movement of "flow" in spatially extended systems (see, e.g., [51]). Consider random processes $\mathbf{X}$ and $\mathbf{Y}$ observed at different spatial locations in a dynamical system. It is argued in [51] that a peak in the mutual information for $\tau > 0$ implies information is moving from the random process $\mathbf{X}$ to the random process $\mathbf{Y}$. For some systems this is certainly true. For example, a propagating pulse along a spatial dimension (similar to that used in our radar example) would undoubtedly lend itself to this interpretation. However, it has been argued that a better measure of information transfer is to quantify the amount of information one random process carries about the *dynamics* of the other. To quantify such dynamic dependency, the transfer entropy was introduced by [47] and incorporates a dynamic model of the random processes directly in how coupling is defined.

In the following discussion we will require a slight change in our notation. To this point we have discussed different random processes $\mathbf{X}$ and $\mathbf{Y}$. This notation becomes cumbersome, however, when more than two random processes are being considered. In our Transfer Entropy examples, we will consider multi-degree freedom systems that produce a number of output signals, each of which modeled as a different random process. In what follows we therefore assume that we have observed the signals $x_i(t_n)$, $i = 1 \cdots M$ as the output of an $M-$degree-of-freedom dynamical system and that we have measured these signals at particular times t_n, $n = 1 \cdots N$. As we have done throughout this text, we choose to model each sampled value $x_i(t_n)$ as a a continuos random variable X_{in} with probability density $p_{X_{in}}(x_i(t_n))$. The vector of random variables $\mathbf{X}_i \equiv (X_{i1}, X_{i2}, \cdots, X_{iN})$ defines the ith random process and will be used to model the sequence of observations $\mathbf{x}_i \equiv x_i(t_n)$, $n = 1 \cdots N$. Using this notation, we can also define the joint PDF $p_{\mathbf{X}_i}(\mathbf{x}_i)$ which specifies the probability of observing such a sequence.

Recall from Chapter 1 that a Markov model is a probabilistic model that uses conditional probability to describe the dynamics of a random process. For example, we might specify $p_{X_{in}}(x_i(t_n)|x_i(t_{n-1}))$ as the probability of observing the value $x_i(t_n)$ given that we have already observed $x_i(t_{n-1})$. The idea that knowledge of past observations changes the likelihood of future observations is certainly common in dynamical systems. A dynamical system whose output is a repeating sequence of $010101 \cdots$ is equally likely to be in state 0 or state 1 (probability 0.5) if the system is observed at a randomly chosen time. However, if we know the value at $t_1 = 0$ the value $t_2 = 1$ is known with probability 1. This concept lies at the heart of the $\mathcal{P}^{th}$ order Markov model, which by definition obeys (see Eqn. 1.57, Chapter 1)

$$p_{X_i}(x_i(t_{n+1})|x_i(t_n), x_i(t_{n-1}), x_i(t_{n-2}), \cdots, x_i(t_{n-\mathcal{P}})) =$$
$$p_{X_i}(x_i(t_{n+1})|x_i(t_n), x_i(t_{n-1}), x_i(t_{n-2}), \cdots, x_i(t_{n-\mathcal{P}}), x_i(t_{n-\mathcal{P}-1}), \cdots)$$
$$\equiv p_{X_i}(x_i(t_n)^{(1)}|\mathbf{x}_i(t_n)^{(\mathcal{P})}) \tag{6.48}$$

That is to say, the probability of the random variable attaining the value $x_i(t_{n+1})$ is conditional on the previous $\mathcal{P}$ values only. In the same way that we have denoted the time lags with a superscript, e.g., $\mathcal{P}$, we similarly denote the unit time step advance, e.g., we write $p_{X_i}(x_i(t_{n+1})) \equiv p_{X_i}(x_i(t_n)^{(1)})$.

Armed with this notation we consider the work of Kaiser and Schreiber [19] and define the continuous transfer entropy between random processes $X_i(t)$ and $X_j(t)$

$$TE_{j \to i}(t_n) = \int_{\mathbb{R}^{\mathcal{P}+Q+1}} p_{X_i}\left(x_i(t_n)^{(1)}|\mathbf{x}_i^{(\mathcal{P})}(t_n), \mathbf{x}_j^{(Q)}(t_n)\right)$$
$$\times \log_2\left(\frac{p_{X_i}(x_i(t_n)^{(1)}|\mathbf{x}_i^{(\mathcal{P})}(t_n), \mathbf{x}_j^{(Q)}(t_n))}{p_{\mathbf{X}_i}(x_i(t_n)^{(1)}|\mathbf{x}_i^{(\mathcal{P})})}\right) dx_i(t_n^{(1)}) d\mathbf{x}_i(t_n)^{(\mathcal{P})} d\mathbf{x}_j(t_n)^{(Q)} \tag{6.49}$$

where the notation $\int_{\mathbb{R}^N}$ is used to denote the N-dimensional integral over the support of the random variables. By definition, this measure quantifies the ability of the random process $\mathbf{X}_j$ to predict the dynamics of the random process $\mathbf{X}_i$. To see why, we can examine the argument of the logarithm. In the event that the dynamics of $\mathbf{X}_i$ are *not* coupled to $\mathbf{X}_j$ one has the Markov model in the denominator of (6.49). However, should $\mathbf{X}_j$ carry information about the transition probabilities of $\mathbf{X}_i$, the numerator is a better model. The transfer entropy is effectively mapping the difference between these hypotheses to the scalar $TE_{j \to i}(t_n)$. Thus, in the same way that mutual information measures deviations from the hypothesis of independence, transfer entropy measures deviations from the hypothesis that the dynamics of $\mathbf{X}_i$ can be described entirely by its own past history and that no new information is gained by considering the dynamics of system $\mathbf{X}_j$.

As with the mutual information the estimation of (6.49) is greatly aided by assuming that the processes are stationary and ergodic, with each observation coming from the same underlying probability distribution. In this case the absolute time index t_n is of no consequence and may be therefore dropped from the notation. With the assumption of stationarity, we may use the law of conditional probability (1.20) and expand Eqn. (6.49) as

$$TE_{j \to i} = \int_{\mathbb{R}^{P+Q+1}} p_{X_i^{(1)} \mathbf{X}_i \mathbf{X}_j} \left(x_i^{(1)}, \mathbf{x}_i^{(P)}, \mathbf{x}_j^{(Q)} \right) \log_2 \left(p_{X_i^{(1)} \mathbf{X}_i \mathbf{X}_j} (x_i^{(1)}, \mathbf{x}_i^{(P)}, \mathbf{x}_j^{(Q)}) \right)$$
$$\times \, dx_i^{(1)} d\mathbf{x}_i^{(P)} d\mathbf{x}_j^{(Q)}$$
$$- \int_{\mathbb{R}^{P+Q}} p_{\mathbf{X}_i \mathbf{X}_j} \left(\mathbf{x}_i^{(P)}, \mathbf{x}_j^{(Q)} \right) \log_2 \left(p_{\mathbf{X}_i \mathbf{X}_j} (\mathbf{x}_i^{(P)}, \mathbf{x}_j^{(Q)}) \right) d\mathbf{x}_i^{(P)} d\mathbf{x}_j^{(Q)}$$
$$- \int_{\mathbb{R}^{P+1}} p_{X_i^{(1)} \mathbf{X}_i} \left(x_i^{(1)}, \mathbf{x}_i^{(P)} \right) \log_2 \left(p_{X_i^{(1)} \mathbf{X}_i} (x_i^{(1)}, \mathbf{x}_i^{(P)}) \right) dx_i^{(1)} d\mathbf{x}_i^{(P)}$$
$$+ \int_{\mathbb{R}^P} p_{\mathbf{X}_i} \left(\mathbf{x}_i^{(P)} \right) \log_2 \left(p_{\mathbf{X}_i} (\mathbf{x}_i^{(P)}) \right) d\mathbf{x}_i^{(P)}$$
$$= -h_{X_i^{(1)} \mathbf{X}_i^{(P)} \mathbf{X}_j^{(Q)}} + h_{\mathbf{X}_i^{(P)} \mathbf{X}_j^{(Q)}} + h_{X_i^{(1)} \mathbf{X}_i^{(P)}} - h_{\mathbf{X}_i^{(P)}} \tag{6.50}$$

where the terms $h_{\mathbf{X}} = -\int_{\mathbb{R}^M} p_{\mathbf{X}}(\mathbf{x}) \log_2 (p(\mathbf{x})) \, d\mathbf{x}$ are again the differential entropies associated with the M-dimensional random variable $\mathbf{X}$. Like the mutual information we see that the transfer entropy is comprised of differential entropies. In what follows we will try to better understand this quantity and then demonstrate its usefulness in several applications.

It is instructive to study this quantity first in the case where an analytical solution is possible. Again assuming that both random processes $\mathbf{X}_i$ and $\mathbf{X}_j$ are jointly Gaussian distributed (see 6.34), we may follow the same procedure as was used to derive (6.35) which ultimately yields

$$TE_{j \to i} = \frac{1}{2} \log_2 \left(\frac{|C_{\mathbf{X}_i^{(P)} \mathbf{X}_j^{(Q)}}||C_{X_i^{(1)} \mathbf{X}_i^{(P)}}|}{|C_{X_i^{(1)} \mathbf{X}_i^{(P)} \mathbf{X}_j^{(Q)}}||C_{\mathbf{X}_i}|} \right). \tag{6.51}$$

For large values of $\mathcal{P}, \mathcal{Q}$ the needed determinants become difficult to compute. We therefore employ a simplification to the model that retains the spirit of the transfer entropy, but that makes an analytical solution more tractable. In our approach, we set $\mathcal{P} = \mathcal{Q} = 1$, i.e., both random processes are assumed to follow a first-order Markov model. However, we allow the time interval between the random processes to vary, just as we did for the mutual information. Specifically, we model $X_i(t)$ as the first-order Markov model $p_{X_i}(x_i(t_n + \Delta_t)|x_i(t_n))$ and use the TE to consider the alternative $p_{X_i}(x_i(t_n + \Delta_t)|x_i(t_n), x_j(t_n + \tau))$ (note that $t_{n+1} = t_n + \Delta_t$). Although we are only using first-order Markov models, by varying the time delay τ we can explore whether or not the random variable $X_j(t_n + \tau)$ carries information about the transition probability $p_{X_i}(x_i(t_n + \Delta_t)|x_i(t_n))$. Should consideration of $x_j(t_n + \tau)$ provide no additional knowledge about the dynamics of $x_i(t_n)$ the transfer entropy will be zero, rising to some positive value should $x_j(t_n + \tau)$ carry information not possessed in $x_j(t_n)$.

In what follows we refer to this particular form of the TE as the time-delayed transfer entropy, or TDTE. In this simplified situation the needed covariance matrices are

$$C_{X_i X_j}(\tau) = \begin{bmatrix} E[(x_i(t_n) - \bar{x}_i)^2] & E[(x_i(t_n) - \bar{x}_i)(x_j(t_n + \tau) - \bar{x}_j)] \\ E[(x_j(t_n + \tau) - \bar{x}_j)(x_i(t_n) - \bar{x}_i)] & E[(x_j(t_n + \tau) - \bar{x}_j)^2] \end{bmatrix}$$

$$C_{X_i^{(1)} X_i X_j}(\tau) = \begin{bmatrix} E[(x_i(t_n + \Delta_t) - \bar{x}_i)^2] & E[(x_i(t_n + \Delta_t) - \bar{x}_i)(x_i(t_n) - \bar{x}_i)] \\ E[(x_i(t_n) - \bar{x}_i)(x_i(t_n + \Delta_t) - \bar{x}_i)] & E[(x_i(t_n) - \bar{x}_i)^2] \\ E[(x_j(t_n + \tau) - \bar{x}_j)(x_i(t_n + \Delta_t) - \bar{x}_i)] & E[(x_j(t_n + \tau) - \bar{x}_j)(x_i(t_n) - \bar{x}_i)] \end{bmatrix}$$

$$\begin{matrix} E[(x_i(t_n + \Delta_t) - \bar{x}_i)(x_j(t_n + \tau) - \bar{x}_j)] \\ E[(x_i(t_n) - \bar{x}_i)(x_j(t_n + \tau) - \bar{x}_j)] \\ E[(x_j(t_n + \tau) - \bar{x}_j)^2] \end{matrix}$$

$$C_{X_i^{(1)} X_i} = \begin{bmatrix} E[(x_i(t_n + \Delta_t) - \bar{x}_i)^2] & E[(x_i(t_n + \Delta_t) - \bar{x}_i)(x_i(t_n) - \bar{x}_i)] \\ E[(x_i(t_n) - \bar{x}_i)(x_i(t_n + \Delta_t) - \bar{x}_i)] & E[(x_i(t_n) - \bar{x}_i)^2] \end{bmatrix}.$$

$$(6.52)$$

and $C_{X_i X_i} = E[(x_i(t_n) - \bar{x}_i)^2] \equiv \sigma_i^2$ is simply the variance of each observation in the random process $\mathbf{X}_i$. The assumption of stationarity also allows us to write $E[(x_i(t_n + \Delta_t) - \bar{x}_i)^2] = \sigma_i^2$ and $E[(x_j(n + \tau) - \bar{x}_j)^2] = \sigma_j^2$. Making these substitutions into (6.51) yields the expression

$$TE_{j \to i}(\tau) =$$

$$\frac{1}{2} \log_2 \left[\frac{\left(1 - \rho_{X_i X_i}^2(\Delta_t)\right)\left(1 - \rho_{X_i X_j}^2(\tau)\right)}{1 - \rho_{X_i X_j}^2(\tau) - \rho_{X_i X_j}^2(\tau - \Delta_t) - \rho_{X_i X_i}^2(\Delta_t) + 2\rho_{X_i X_i}(\Delta_t)\rho_{X_i X_j}(\tau)\rho_{X_i X_j}(\tau - \Delta_t)} \right]$$

$$(6.53)$$

where we have defined particular expectations in the covariance matrices using the shorthand $\rho_{X_i X_j}(\tau) \equiv E[(x_i(t_n) - \bar{x}_i)(x_j(t_n + \tau) - \bar{x}_j)]/\sigma_i \sigma_j$. Note that the covariance matrices are positive-definite matrices and that the determinant of a positive definite matrix is positive [20]. Thus the quantity inside the logarithm will always be positive and the logarithm will exist. In fact, it turns out that by construction the argument of the logarithm in (6.53) is always ≥ 1 so that the TE is always positive.

Now, the hypothesis that the TE was designed to test is whether or not past values of the process X_j carry information about the transition probabilities of the second process X_i. Thus, if we are to keep with the original intent of the measure we would only consider $\tau < 0$. However, this restriction is only necessary if one implicitly assumes a non-zero TE means X_j is *influencing* the transition $p_{X_i}(x_i(t_n + \Delta_t)|x_i(t_n + \tau))$ as opposed to simply carrying additional information *about* the transition. Again, this latter statement is a more accurate depiction of what the TE is really measuring and we have found it useful to consider both negative and positive delays τ in trying to understand coupling among system components.

It is also interesting to note the bounds of this function. Certainly for DC signals (i.e., $x_i(t_n)$, $x_j(t_n)$ constant) we have $\rho_{X_i X_i}(\Delta_t) = \rho_{X_i X_j}(\tau) = 0 \; \forall \; \tau$ and the transfer entropy is zero for any choice of time scales τ defining the Markov processes. Knowledge of X_j does not aid in forecasting X_i simply because the transition probability in going from $x_i(t_n)$ to $x_i(t_n + \Delta_t)$ is always unity. Likewise, if there is no coupling between system components we have $\rho_{X_i X_j}(\tau) = 0$ and the TDTE becomes $TE_{j \to i}(\tau) = \frac{1}{2} \log_2 \left[\frac{1 - \rho_{X_i X_i}^2(\Delta_t)}{1 - \rho_{X_i X_i}^2(\Delta_t)} \right] = 0$.

At the other extreme, for *perfectly* coupled systems, i.e., $X_i = X_j$, consider $\tau \to 0$. In this case, we have $\rho^2_{X_i X_j}(\tau) \to 1$, and $\rho_{X_i X_j}(\tau - \Delta_t) \to \rho_{X_i X_i}(-\Delta_t) = \rho_{X_i X_i}(\Delta_t)$ (in this last expression we have noted the symmetry of the function $\rho_{X_i X_i}(\tau)$ w.r.t. the time delay). The transfer entropy then becomes

$$TE_{j \to i}(0) = \frac{1}{2} \log_2 \begin{bmatrix} 0 \\ 0 \end{bmatrix} \to 0 \qquad (6.54)$$

and the random process X_j at $\tau = 0$ is seen to carry no *additional* information about the dynamics of X_i simply due to the fact that in this special case we have $p_{X_i}(x_i(t_n + \Delta_t)|x_i(t_n)) = p_{X_i}(x_i(t_n + \Delta_t)|x_i(t_n), x_i(t_n))$. These extremes highlight the care that must be taken in interpreting the transfer entropy. Because the TDTE is zero for both the perfectly coupled and uncoupled cases, it must not interpret the measure to quantify the coupling strength between two random processes. Rather, the TDTE measures the additional information provided by one random process about the dynamics of another.

We should also point out that the average mutual information function can resolve this ambiguity if one wishes to use the TDTE as a diagnostic of dynamical coupling. For two Gaussian random processes the mutual information given by (6.38) is known to be $I_{X_i X_j}(\tau) = -\frac{1}{2} \log_2 \left[1 - \rho^2_{X_i X_j}(\tau) \right]$. Hence, for perfect coupling $I_{X_i X_j}(0) \to \infty$, whereas for uncoupled systems $I_{X_i X_j}(0) \to 0$. Computing both time-delayed mutual information and transfer entropies can therefore permit stronger inference about dynamical coupling.

6.3.1 Transfer Entropy for Second-Order Linear Systems

In this section we apply the transfer entropy to the study of coupling among components of a driven system. To fully define the TDTE for Gaussian random processes, the auto- and cross-correlation functions $\rho_{XX}(\tau)$, $\rho_{XY}(\tau)$ are required. They are derived here for a general class of linear system found frequently in the modeling and analysis of physical processes. Consider the system defined by the general expression

$$\mathbf{M\ddot{x}}(t) + \mathbf{C\dot{x}}(t) + \mathbf{Kx}(t) = \mathbf{f}(t) \qquad (6.55)$$

where $\mathbf{x}(t) \equiv (x_1(t), x_2(t), \cdots, x_M(t))^T$ is the system's response to the forcing function(s) $\mathbf{f}(t) \equiv (f_1(t), f_2(t), \cdots, f_M(t))^T$, and $\mathbf{M}$, $\mathbf{C}$, $\mathbf{K}$ are $M \times M$ constant coefficient matrices that capture the system's physical properties. Thus, we are considering second-order, $M-$degree-of-freedom (DOF), linear systems. It is assumed that we may measure the response of this system at any of the DOFs and/or the forcing functions. One physical embodiment of this system is shown schematically in Figure 6.7. Five masses are coupled together via restoring elements k_i (springs) and dissipative elements, c_i (dash-pots). The first mass is fixed to a boundary while the driving force is applied at the end mass. If the response data $\mathbf{x}(t)$ are each modeled as a stationary random

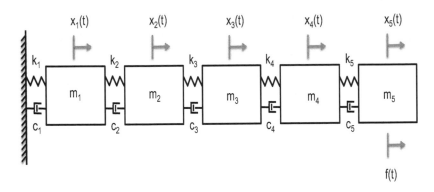

FIGURE 6.7
Physical system modeled by Eqn. (6.55). Here, an $M = 5$ degree-of-freedom structure is represented by masses coupled together via both restoring and dissipative elements. Forcing is applied at the end mass.

process, we may use the mutual information and transfer entropy to answer questions about the coupling between any two masses. We can explore this relationship as a function of coupling strength and also which particular mass response data we choose to analyze.

We require a general expression for the cross-correlation between any two DOFs i, $j \in [1, M]$. In other words, we require the expectation $E[x_i(n)x_j(n + \tau)]$ for any combination of i, j. Such an expression can be obtained by first transforming coordinates, i.e., let $\boldsymbol{x}(t) = \mathbf{u}\boldsymbol{\eta}(t)$ where the matrix $\mathbf{u}$ contain the non-trivial solutions to the eigenvalue problem $|\mathbf{M}^{-1}\mathbf{K} - \omega_i^2\mathbf{I}|\mathbf{u}_i = 0$ as its columns [30]. Here the eigenvalues are the natural frequencies of the system, denoted ω_i, $i = 1 \cdots M$. Making the above coordinate transformation, substituting into (6.55) and then pre-multiplying both sides by $\mathbf{u}^T$ allows the equations of motion to be uncoupled and written separately as

$$\ddot{\eta}_i(t) + 2\zeta_i\omega_i\dot{\eta}_i(t) + \omega_i^2\eta_i(t) = \mathbf{u}_i^T\mathbf{f}(t) \equiv q_i(t). \qquad (6.56)$$

where the eigenvectors have been normalized such that $\mathbf{u}^T\mathbf{M}\mathbf{u} = \mathbf{I}$ (the identity matrix). In the above formulation we have also made the assumption that $\mathbf{C} = \alpha\mathbf{K}$, i.e., the dissipative coupling $\mathbf{C}\dot{\mathbf{x}}(t)$ is of the same form as the restoring term, albeit scaled by the constant $\alpha << 1$ (i.e., a lightly damped system). To obtain the form shown in (6.56), we introduce the dimensionless damping coefficient $\zeta_i = \frac{\alpha}{2}\omega_i$.

The general solution to these uncoupled, linear equations is well-known [30] and can be written as the convolution

$$\eta_i(t) = \int_0^\infty h_i(\theta)q_i(t - \theta)d\theta \qquad (6.57)$$

where $h_i(\theta)$ is the impulse response function

$$h_i(\theta) = \frac{1}{\omega_{di}} e^{-\zeta_i \omega_i \theta} \sin(\omega_{di}\theta) \tag{6.58}$$

and $\omega_{di} \equiv \omega_i \sqrt{1 - \zeta_i^2}$. In general terms, we therefore have

$$x_i(t) = \sum_{l=1}^{M} u_{il}\eta_l(t)$$

$$= \int_0^\infty \sum_{l=1}^{M} u_{il}h_l(\theta)q_l(t - \theta)d\theta \tag{6.59}$$

If we further consider the excitation $\mathbf{f}(t)$ to be a zero-mean random process, so too will be $q_l(t)$. Using this model, we may construct the covariance

$$E[x_i(t)x_j(t + \tau)] =$$

$$E\left[\int_0^\infty \int_0^\infty \sum_{l=1}^{M} \sum_{m=1}^{M} u_{il}u_{jm}h_l(\theta_1)h_m(\theta_2)q_l(t - \theta_1)q_m(t + \tau - \theta_2)d\theta_1 d\theta_2\right]$$

$$= \int_0^\infty \int_0^\infty \sum_{l=1}^{M} \sum_{m=1}^{M} u_{il}u_{jm}h_l(\theta_1)h_m(\theta_2)E[q_l(t - \theta_1)q_m(t + \tau - \theta_2)]d\theta_1 d\theta_2$$

$$\tag{6.60}$$

which is a function of the mode shapes $\mathbf{u}_i$, the impulse response function $h_i(\cdot)$ and the covariance of the modal forcing matrix. Knowledge of this covariance matrix can be obtained from knowledge of the forcing covariance matrix $R_{F_l F_m}(\tau) \equiv E[f_l(t)f_m(t + \tau)]$. Recalling that

$$q_l(t) = \sum_{p=1}^{M} u_{lp}f_p(t) \tag{6.61}$$

we write

$$E[q_l(t - \theta_1)q_m(t + \tau - \theta_2)] = \sum_{p=1}^{M} \sum_{q=1}^{M} u_{lq}u_{mp}E[f_q(t - \theta_1)f_p(t + \tau - \theta_2)]. \tag{6.62}$$

It is assumed that the random vibration inputs are uncorrelated, i.e., $E[f_q(t)f_p(t)] = 0 \ \forall \ q \neq p$ in which case the above can be simplified as

$$E[q_l(t - \theta_1)q_m(t + \tau - \theta_2)] = \sum_{p=1}^{M} u_{lp}u_{mp}E[f_p(t - \theta_1)f_p(t + \tau - \theta_2)]. \tag{6.63}$$

The most common linear models assume the input is applied at a single DOF, i.e., $f_p(t)$ is non-zero only for $p = P$. For a load applied at DOF P, the auto-covariance becomes

$$
E[x_i(t)x_j(t+\tau)] = \int_0^\infty \int_0^\infty \sum_{l=1}^M \sum_{m=1}^M u_{il}u_{jm}u_{lP}u_{mP}h_l(\theta_1)h_m(\theta_2)
$$

$$
\times E[f_P(t-\theta_1)f_P(t+\tau-\theta_2)]d\theta_1 d\theta_2
$$

$$
= \sum_{l=1}^M \sum_{m=1}^M u_{il}u_{jm}u_{lP}u_{mP} \int_0^\infty h_l(\theta_1) \int_0^\infty h_m(\theta_2)
$$

$$
\times E[f_P(t-\theta_1)f_P(t+\tau-\theta_2)]d\theta_2 d\theta_1. \quad (6.64)
$$

The inner integral can be further evaluated as

$$
\int_0^\infty h_m(\theta_2)E[f_P(t-\theta_1)f_P(t+\tau-\theta_2)]d\theta_2
$$

$$
= \int_0^\infty h_m(\theta_2) \int_{-\infty}^\infty S_{FF}(\omega)e^{i\omega(\tau-\theta_2+\theta_1)}d\omega d\omega_2. \quad (6.65)
$$

where we have written the forcing covariance as the inverse Fourier transform of the power spectral density function, $S_{FF}(\omega)$ [40]. We have already assumed the forcing is comprised of statistically independent values (i.e., the process is i.i.d.), in which case the forcing power spectral density $S_{FF}(\omega) = const \; \forall \omega$. Denoting this constant $S_{FF}(0)$, we note that the Fourier Transform of a constant is simply $\int_{-\infty}^\infty S_{FF}(0) \times e^{i\omega t}dt = S_{FF}(0) \times \delta(t)$, hence our integral becomes

$$
\int_0^\infty h_m(\theta_2)E[f_P(t-\theta_1)f_P(t+\tau-\theta_2)]d\theta_2
$$

$$
= \int_0^\infty h_m(\theta_2)S_{FF}(0)\delta(\tau-\theta_2+\theta_1) = h(\tau+\theta_1)S_{FF}(0). \quad (6.66)
$$

Returning to Eqn. (6.64) we have

$$
E[x_i(t)x_j(t+\tau)] = \int_0^t \sum_{l=1}^M \sum_{m=1}^M u_{lP}u_{mP}u_{il}u_{jm}h_l(\theta_1)h_m(\theta_1+\tau)S_{FF}(0)d\theta_1.
$$

$$
(6.67)
$$

At this point we can simplify the expression by carrying out the integral. Substituting the expression for the impulse response (6.58), the needed expectation (6.67) becomes [3, 9]

$$
R_{X_iX_j}(\tau) = \frac{S_{FF}(0)}{4} \sum_{l=1}^M \sum_{m=1}^M u_{lP}u_{mP}u_{il}u_{jm} \left[A_{lm}e^{-\zeta_m\omega_m\tau}\cos(\omega_{dm}\tau) \right.
$$

$$
\left. + B_{lm}e^{-\zeta_m\omega_m\tau}\sin(\omega_{dm}\tau) \right] \quad (6.68)
$$

where

$$A_{lm} = \frac{8\left(\omega_l\zeta_l + \omega_m\zeta_m\right)}{\omega_l^4 + \omega_m^4 + 4\omega_l^3\omega_m\zeta_l\zeta_m + 4\omega_m^3\omega_l\zeta_l\zeta_m + 2\omega_m^2\omega_l^2\left(-1 + 2\zeta_l^2 + 2\zeta_m^2\right)}$$

$$B_{lm} = \frac{4\left(\omega_l^2 + 2\omega_l\omega_m\zeta_l\zeta_m + \omega_m^2\left(-1 + 2\zeta_m^2\right)\right)}{\omega_{dm}\left(\omega_l^4 + \omega_m^4 + 4\omega_l^3\omega_m\zeta_l\zeta_m + 4\omega_m^3\omega_l\zeta_l\zeta_m + 2\omega_m^2\omega_l^2\left(-1 + 2\zeta_l^2 + 2\zeta_m^2\right)\right)}$$

$$(6.69)$$

We can further normalize this function to give

$$\rho_{X_iX_j}(\tau) = R_{X_iX_j}(\tau)/\sqrt{R_{X_iX_i}(0)R_{X_jX_j}(0)} \tag{6.70}$$

for the normalized auto- and cross-correlation functions. The expression (6.70) is sufficient to define both the time-delayed mutual information function (6.38) and the time-delayed transfer entropy (6.53) between two joint normally distributed random variables.

It will also prove instructive to study the TDTE between the drive and response. This requires $R_{X_iF_P}(\tau) \equiv E[x_i(t)f_P(t+\tau)]$. Following the same procedure as the above results in the expression

$$R_{X_iF_P}(\tau) = \begin{cases} \mathbf{S}_{FF}(0)\sum_{m=1}^{M} u_{im}u_{mP}h_m(-\tau) & : \quad \tau \leq 0 \\ 0 & : \quad \tau > 0 \end{cases} . \tag{6.71}$$

Normalizing by the product of variances of X_i and F_P yields the normalized correlation function, $\rho_{if}(\tau)$. This expression may be substituted into the expression for the transfer entropy to yield the TDTE between drive and response. At this point we have completely defined the analytical TDTE for a broad class of second order linear systems in terms of the auto- and cross-correlation functions In this special case, we can derive an alternative estimator to those described in Section 6.1.

Assume we have recorded the signals $x_i(n\Delta_t)$, $x_j(n\Delta_t)$ $n = 1\cdots N$ with a fixed sampling interval Δ_t. In order to estimate the TDTE we require a means of estimating the normalized correlation functions $\rho_{X_iX_j}(\tau)$ which can be substituted into Eqn. (6.53). While different estimators of correlation functions exist (see e.g., [4]), we use a frequency domain estimator. This estimator relies on the assumption that the observed data are the output of a stationary random process, an assumption we have already made. For this type of data, the Wiener-Khinchin Theorem tells us that the cross-spectral density and cross-covariance functions are related via Fourier transform as

$$\int_{-\infty}^{\infty} E[x_i(t)x_j(t+\tau)]e^{-i2\pi f\tau}d\tau = S_{X_jX_i}(f) \equiv \lim_{T\to\infty} E\left[\frac{X^*(f)Y(f)}{2T}\right].$$

$$(6.72)$$

One approach is to estimate the spectral density $\hat{S}_{X_jX_i}(f)$ and then inverse Fourier transform to give $\hat{R}_{X_iX_j}(\tau)$. We further rely on the ergodic theorem

(6.20) which (when applied to probability) allows one to write expectations defined over multiple realizations to be well-approximated temporally averaging over a finite number of samples. Working with discrete time indices, we divide the observations $x_i(n)$, $x_j(n)$, $n = 1 \cdots N$ into S segments of length N_s, (possibly) overlapping by L points. Taking the discrete Fourier transform of each segment, e.g., $X_{is}(k) = \sum_{n=0}^{N_s-1} x_i(n + sN_s - L)e^{-i2\pi kn/N_s}$, $s = 0 \cdots S - 1$ and averaging gives the estimator

$$\hat{S}_{X_j X_i}(k) = \frac{\Delta_t}{N_s S} \sum_{s=0}^{S-1} \hat{X}_{is}^*(k)\hat{X}_{js}(k) \tag{6.73}$$

at discrete frequency k (denotes complex conjugate). This quantity is then inverse discrete Fourier transformed to give

$$\hat{R}_{X_i X_j}(n) = \frac{1}{N_s} \sum_{k=0}^{N_s-1} \hat{S}_{X_j X_i}(k)e^{i2\pi kn/N_s}. \tag{6.74}$$

Finally, we may normalize the estimate to give the cross-correlation coefficient

$$\hat{\rho}_{X_i X_j}(n) = \hat{R}_{X_i X_j}(n)/\sqrt{\hat{R}_{X_i X_i}(0)\hat{R}_{X_j X_j}(0)}. \tag{6.75}$$

This estimator is asymptotically consistent and unbiased and can therefore be substituted into Eqn. (6.53) to produce very accurate estimates of the TE (see examples to follow). In the general (nonlinear) case, kernel density estimators are typically used and are known to be poor in many cases, particularly when data are scarce (see e.g., [15, 39]).

As an example, we consider the five DOF systems governed by Eqn. (6.55) where

$$\mathbf{M} = \begin{bmatrix} m_1 & 0 & 0 & 0 & 0 \\ 0 & m_2 & 0 & 0 & 0 \\ 0 & 0 & m_3 & 0 & 0 \\ 0 & 0 & 0 & m_4 & 0 \\ 0 & 0 & 0 & 0 & m_5 \end{bmatrix}$$

$$\mathbf{C} = \begin{bmatrix} c_1 + c_2 & -c_2 & 0 & 0 & 0 \\ -c_2 & c_2 + c_3 & -c_3 & 0 & 0 \\ 0 & -c_3 & c_3 + c_4 & -c_4 & 0 \\ 0 & 0 & -c_4 & c_4 + c_5 & -c_5 \\ 0 & 0 & 0 & -c_5 & c_5 \end{bmatrix}$$

$$\mathbf{K} = \begin{bmatrix} k_1 + k_2 & -k_2 & 0 & 0 & 0 \\ -k_2 & k_2 + k_3 & -k_3 & 0 & 0 \\ 0 & -k_3 & k_3 + k_4 & -k_4 & 0 \\ 0 & 0 & -k_4 & k_4 + k_5 & -k_5 \\ 0 & 0 & 0 & -k_5 & k_5 \end{bmatrix} \tag{6.76}$$

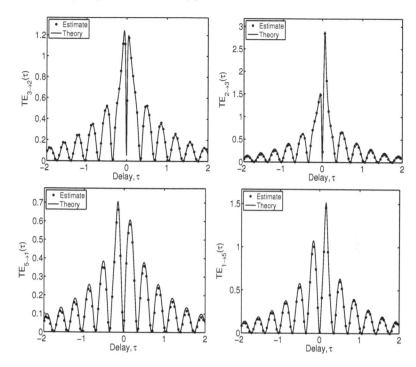

FIGURE 6.8

Time-delay transfer entropy between masses 2 and 3 (top row) and 1 and 5 (bottom row) of a five-degree of freedom system driven at mass $P = 5$.

are constant coefficient matrices of a form commonly used to describe structural systems. In this case, these particular matrices describe the motion of a cantilevered structure where we assume a joint normally distributed random process applied at the end mass, i.e., $\mathbf{f}(t) = (0, 0, 0, 0, \mathcal{N}(0, 1))$. In this first example we examine the TDTE between response data collected from two different points on the structure. We fix $m_i = 0.01 \ kg$, $c_i = 0.1 \ N \cdot s/m$, and $k_i = 10 \ N/m$ for each of the $i = 1 \cdots 5$ degrees of freedom (thus we are using $\alpha = 0.01$ in the modal damping model $\mathbf{C} = \alpha \mathbf{K}$). The system response data $x_i(n\Delta_t)$, $n = 1 \cdots 2^{15}$ to the stochastic forcing are then generated via numerical integration. For simulation purposes we used a time step of $\Delta_t = 0.01 \ s$ which is sufficient to capture all five of the system modes (the lowest of which resides at a frequency of $\omega_1 = 9.00 \ rad/s$). Based on these parameters, we generated the analytical expressions $TE_{3 \to 2}(\tau)$ and $TE_{2 \to 3}(\tau)$ and also $TE_{5 \to 1}(\tau)$ and $TE_{1 \to 5}(\tau)$ for illustrative purposes. These are shown in Figure 6.8 along with the estimates formed using the Fourier transform-based procedure.

With Figure 6.8 in mind, first consider negative delays only where $\tau < 0$. Clearly, the further the random variable $X_j(n + \tau)$ is from $X_i(n)$, the less information it carries about the probability of X_i transitioning to a new state

Δ_t seconds into the future. This is to be expected from a stochastically driven system and accounts for the decay of the transfer entropy to zero for large $|\tau|$. However, we also see periodic returns to the point $TE_{j \to i}(\tau) = 0$ for even small temporal separation. Clearly this is a reflection of the periodicity observed in second-order linear systems. In fact, for this system the dominant period of oscillation is $2\pi/\omega_1 = 0.698$ seconds. It can be seen that the argument of the logarithm in (6.53) periodically reaches a minimum value of unity at precisely half this period, thus we observe zeros of the TDTE at times $(i-1) \times \pi/\omega_1$, $i = 1 \cdots$. In this case the TDTE is going to zero *not* because the random variables $X_j(n+\tau)$, $X_i(n)$ are unrelated, but because knowledge of one allows us to exactly predict the position of the other (no additional information is present). We believe this is likely to be a feature of most systems possessing an underlying periodicity and is one reason why using the TE as a measure of coupling must be done with care.

We also point out that values of the TDTE are non-zero for positive delays as well. Again, so long as we interpret the TE as a measure of predictive power this makes sense. That is to say, future values X_j can aid in predicting the current dynamics of X_i. Interestingly, the asymmetry in the TE peaks near $\tau = 0$ may provide the largest clue as to the location of the forcing signal. Consistently we have found that the TE is larger for negative delays when mass closest the driven end plays the role of X_j; conversely, it is larger for positive delays when the mass furthest from the driven end plays this role. So long as the coupling is bi-directional, results such as those shown in Figure 6.8 can be expected in general.

However, the situation is quite different if we consider the case of uni-directional coupling. For example, we may consider $TE_{f \to i}(\tau)$, i.e., the TDTE between the forcing signal and response variable i. This is a particularly interesting case as, unlike in previous examples, there is no feedback from DOF i to the driving signal. Figure 6.9 shows the TDTE between drive and response and clearly highlights the directional nature of the coupling. Past values of the forcing function clearly help in predicting the dynamics of the response. Conversely, future values of the forcing say nothing about transition probabilities for the mass response simply because the mass has not "seen" that information yet. Thus, for uni-directional coupling, the TDTE can easily diagnose whether X_j is driving X_i or vice versa. It can also be noticed from these plots that the drive signal is not that much help in predicting the response as the TDTE is much smaller in magnitude than when computed between masses. We interpret this to mean that the response data are dominated by the physics of the structure (e.g., the structural modes), which is information not carried in the drive signal. Hence, the drive signal offers little in the way of additional predictive power. While the drive signal puts energy into the system, it is not very good at predicting the response. It should also be pointed out that the kernel density estimation techniques are not able to capture these small values of the TDTE. The error in such estimates is larger than these subtle fluctu-

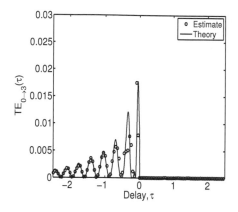

FIGURE 6.9
Time-delay transfer entropy between the forcing and mass 3 for the same five-degree of freedom system driven at mass $P = 5$. The plot is consistent with the interpretation of information moving from the forcing to mass 3.

ations. Only the "linearized" estimator is able to capture the fluctuations in the TDTE for small $(O(10^{-2}))$ values.

It has been suggested that the main utility of the TE is to, given a sequence of observations, assess the direction of information flow in a coupled system. More specifically, one computes the difference $TE_{i \to j} - TE_{j \to i}$ with a positive difference suggesting information flow from i to j (negative differences indicating the opposite) [2, 19]. In the system modeled by (6.55) one would heuristically understand the information as flowing from the drive signal to the response. This is certainly reinforced by Figure 6.9. However, by extension it might seem probable that information would similarly flow from the mass closest the drive signal to the mass closest the boundary (e.g., DOF 5 to DOF 1). We test this hypothesis as a function of the coupling strength between masses. Fixing each stiffness and damping coefficient to the previously used values, we vary k_3 from 1 N/m to 40 N/m and examine the quantity $TE_{i \to j} - TE_{j \to i}$ evaluated at τ^*, defined as the delay at which the TDTE reaches its maximum. Varying k_3 slightly alters the dominant period of the response. By accounting for this shift we eliminate the possibility of capturing the TE at one of its nulls (see Figure 6.8). For example, in Figure 6.8 $\tau^* = -0.15$. Figure 6.10 shows the difference in TDTE as a function of the coupling strength. The result is non-intuitive if one assumes information would move from the driven end toward the non-driven end of the system. For certain DOFs this interpretation holds, for others, it does not. Herein lies the difficulty in interpreting the TE when bi-directional coupling exists.

Rather than being viewed as a measure of information flow, we find it more useful to interpret the difference measure as simply one of predictive power. That is to say, does knowledge of system j help predict system i more

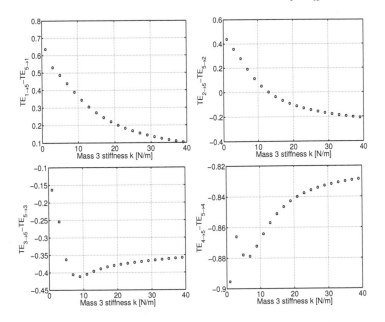

FIGURE 6.10

Difference in time-delay transfer entropy between the driven mass 5 and each other DOF as a function of k_3. A positive difference indicates $TE_{i \to j} > TE_{j \to i}$ and is commonly used to indicate that information is moving from mass i to mass j. Based on this interpretation, negative values indicate information moving from the driven end to the base; positive values indicate the opposite. Even for this linear system, choosing different masses in the analysis can produce very different results. In fact, $TE_{2 \to 5} - TE_{5 \to 2}$ implies a different direction of information transfer depending on the strength of the coupling k_3 (right).

so than i helps predict j. This is a slightly different question. Our analysis suggests that if X_i and X_j are both near the driven end but with DOF i the closer of the two, then knowledge of $\mathbf{X}_j$ is of more use in predicting $\mathbf{X}_i$ than vice versa. This interpretation also happens to be consistent with the notion of information moving from the driven end toward the base. However, as i and j become de-coupled (physically separated) it appears the reverse is true. The random process $\mathbf{X}_i$ is better at predicting $\mathbf{X}_j$ than $\mathbf{X}_j$ is in predicting $\mathbf{X}_i$. Thus, for certain pairs of masses information seems to be traveling from the base toward the drive. One possible explanation is that because the mass $\mathbf{X}_i$ is further removed from the drive signal it is strongly influenced by the vibration of each of the other masses. By contrast, a mass near the driven end is strongly influenced only by the drive signal. Because the dynamics $\mathbf{X}_i$ are influenced heavily by the structure (as opposed to the drive), $\mathbf{X}_i$ does a good job in helping to predict the dynamics everywhere. The main point of this

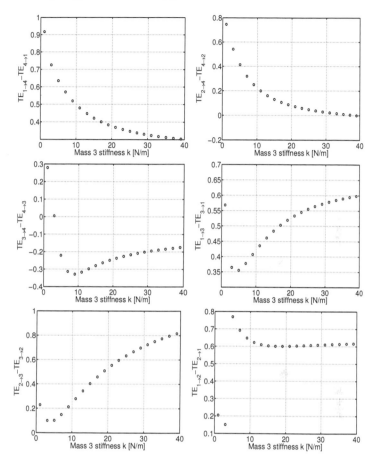

FIGURE 6.11

Difference in TDTE among different combinations of masses. By the traditional interpretation of TE, negative values indicate information moving from the driven end to the base; positive values indicate the opposite.

analysis is that the difference in TE is not at all an unambiguous measure of the direction of information flow.

To further explore this question, we have repeated this numerical experiment for all possible combinations of masses. These results are displayed in Figure 6.11. The same basic phenomenology is observed. If both masses being analyzed are near the driven end, it appears the driven mass $\mathbf{X}_j$ is a better predictor of $\mathbf{X}_i$ for $i < j$. However again, as i and j become decoupled the reverse is true. Our interpretation is that the further the process is removed from the drive signal, the more it is dominated by the other mass dynamics and the boundary conditions. Because such a process is strongly influenced by the other DOFs, it can successfully predict the motion for these other DOFs.

Regardless of how one interprets these results one point should be clear. Simply computing a difference in transfer entropy as an indicator of information flow is most likely a poor way to understand coupling. Even for this simple linear system this interpretation breaks down completely. We would imagine a similar result would hold for more complex systems; however, such systems are beyond our ability to develop analytical expressions. Perhaps a better interpretation is to simply understand the difference in TE as an indicator of which system component provides the most predictive power about the rest of the system dynamics.

In this section we have derived an analytical expression for the transfer entropy for a broad class of second-order linear systems driven by a jointly Gaussian, i.i.d. input. The solution has proven useful both in assessing the quality of one particular estimator of the linearized transfer entropy, and in interpreting the TDTE measure. When the coupling is uni-directional, we have found the TDTE to be an unambiguous indicator of the direction of information flow in a system. However, for bi-directional coupling the situation is significantly more complicated, even for linear systems. We have found that a heuristic understanding of information flow is not always accurate. For example, one might expect information to travel from the driven end of a system toward the non-driven end. In fact, we have shown precisely the opposite to be true. It would seem a safer interpretation is that a positive difference in the transfer entropy tells the practitioner which random process has the greatest predictive power in describing the rest of the system dynamics.

6.3.2   Nonlinearity Detection

We have just discussed the performance of the transfer entropy as a means of understanding linear coupling. We considered this special case in order to get a better understanding of the measure and how it behaves in a known case; however, the true power of the transfer entropy is a measure of *nonlinear* coupling among system components. In fact, the transfer entropy can be used to great effect as a way of distinguishing linear from nonlinear coupling. Several studies (see, e.g., [39] and [42]) have demonstrated that the transfer entropy is more sensitive to the presence of nonlinear coupling in structural systems than is the mutual information function. We consider one such application here.

Assume a two degrees of freedom spring-mass system of the same basic form as in the previous example. We first consider the linear case where the system is governed by Eqn. (6.55) with $k_1 = k_2 = 1.0$ N/m, $c_1 = c_2 = 0.01$ N·s/m, and $m_1 = m_2 = 0.01$ kg. The natural frequencies for the system are $\omega_1 = 6.18, \omega_2 = 16.18$ rad/s. Assuming a proportional damping model, the dimensionless damping ratios become $\zeta_1 = 0.0275, \zeta_2 = 0.0728$ and we can use Eqn. (6.68) to produce an analytical expression for the TDTE between masses. Using these same parameters, the system described by Eqns. (6.55) was also simulated using a fifth-order Runge-Kutta scheme with a time step

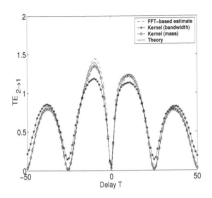
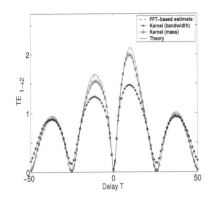

FIGURE 6.12
Different estimates of the time-delayed transfer entropy plotted against theoretical predictions (solid line) for $TE_{2\to1}$ and $TE_{1\to2}$. Linearized estimates obtained via FFT (dashed), kernel estimates using a fixed bandwidth (star), and kernel estimates using a fixed mass (diamond).

of $\Delta_t = 0.02s$. giving time series for the two displacements $x_i(t_n), x_j(t_n)$, $n = 1 \cdots N$ where N was chosen to be $N = 50{,}000$ points.

We estimate the transfer entropy using several estimators discussed above. In particular, we used the aforementioned Fourier Transform approach (only appropriate for linear systems) and also the kernel-density estimation approach given by Eqn. (6.24) with both fixed bandwidth (6.28) and fixed mass (6.31) kernels. Figure 6.12 displays the theoretical TDTE along with the major estimation approaches as a function of integer delay T. Again, the FFT-based approach is only valid in this linear example but produces good results as expected. What we are really attempting to assess here is which approach to use in the general, nonlinear case. We have found that only the fixed-mass kernel can consistently provide an estimate that is in close agreement with theory. The results can be rather sensitive to the number of points M used in the estimator; however, we have found it to be the best non-parametric approach available and therefore use this estimator in what follows.

In order to explore the influence of nonlinearity on the TDTE, we gradually change the coupling between the two masses from linear to nonlinear. The form of the nonlinearity is a bilinear stiffness function, i.e., the linear stiffness k_2 is replaced by a bi-linear stiffness term so that the restoring force assumes the values

$$k_2(x_j - x_i) = \begin{cases} k_2(x_j - x_i) & : & (x_j - x_i) \geq 0 \\ \delta k_2(x_j - x_i) & : & (x_j - x_i) < 0 \end{cases}$$

where the parameter $0 < \delta \leq 1$ controls the degree to which the stiffness is decreased. The discontinuity was implemented numerically by utilizing Henon's integration scheme [16]. Using this approach the change in stiffness could be implemented at exactly $x_j - x_i = 0$ (to the error in the Runge-Kutta algo-

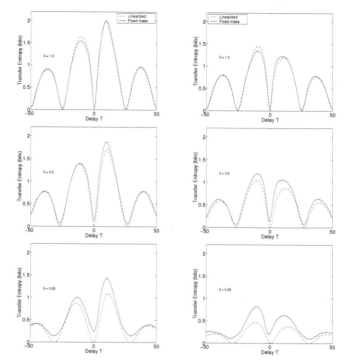

FIGURE 6.13
Time-delayed transfer entropy as estimated using the fixed mass approach
($M = 10$) and the linearized transfer entropy for increasing levels of nonlinearity. Both estimates are plotted as a function of the discrete delay index
T. As the degree of nonlinearity increases the fixed mass estimator begins to
capture statistical properties in the response that the linear estimator cannot.

rithm). We then use both the "linearized" TDTE, formed using the FFT-based
estimator of the previous section, and the non-parametric fixed-mass estimator ($M = 10$ with spherical volume elements) and look for differences in the
resulting curves. By definition the linearized version will reflect only covariance information while the fixed mass estimator is general and should capture
any other order correlations in the response.

The results are shown in Figure 6.13. Notice how the transfer entropy peaks
begin to take on a distorted shape as the degree of nonlinearity is increased.
A characteristic of the TDTE appears to be wider, more skewed, and in most
cases larger peaks in the function. Furthermore, as the nonlinearity increases,
the transfer entropy exhibits larger values relative to that predicted by the
linearized transfer entropy. For example, the maximum peak of the linearized
transfer entropy decreases from a value of 2.0 bits in the linear case ($\delta = 1$) and
takes a value of 1.1 bits in the $\delta = 0.25$ case. By contrast, the transfer entropy
obtained through the kernel-based procedure shows only a slight decrease from

2.0 bits to around 1.4 bits as the degree of nonlinearity goes from $\delta = 0$ to $\delta = 0.25$.

The more general estimator is capturing the influence of the nonlinear coupling in a way that the linearized measure, by definition, cannot. This difference becomes more pronounced as the degree of nonlinearity increases. While there are numerous ways to quantify this difference and use it as a nonlinearity detection statistic, this example has shown the basic utility of the TDTE in detecting the presence of nonlinear coupling among system components.

6.3.2.1 Method of Surrogate Data

One obvious critique of the above analysis is that we are comparing two very different estimators of the TE. Although it is obvious that the primary difference in TDTE curves is the nonlinearity, we cannot rule out differences in the estimators themselves as causing some of the discrepancy. A more sound approach is to use the same estimator of the TDTE, but perform the estimate for both the original signal *and* a "linearized" version of the signal. This is the so-called method of "surrogate data" [43] and is a popular approach for detecting the presence of nonlinearity in dynamical systems.

To be more specific, consider a linear system driven by a stationary random process, $f(t)$. Let us further specify that this random process is described completely by the auto-covariance, $R_{FF}(\tau)$, and marginal distribution $p_F(f)$. This is a fairly general input model, encompassing a wide variety of naturally occuring processes. Now, let us assume that we measure the output of a linear, time-invariant system $x(t) = \int h(t-\tau)f(\tau)d\tau$. It is not hard to show that the output will also be a random process where the joint statistical properties are described entirely by the marginal probability density function, $p_X(x)$, and the auto-correlation $R_{XX}(\tau)$ [43]. If the structure is *nonlinear*, however, there will likely be additional joint moments introduced, e.g., $E[x(t)x(t-\tau_1)x(t-\tau_2)] \neq 0 \forall \tau_1, \tau_2 > 0$. Thus, one could envision estimating this third, joint moment from a sampled response $x(t_n)$ and testing to see if $C_{XXX}(\tau_1, \tau_2) > 0$.

Indeed, this approach is commonly used and forms the basis of higher-order spectral analysis [40]. However, we know that even if the structure is linear there will be some non-zero values due to errors in the estimation process. What we really are testing is therefore $C_{XXX}(\tau_1, \tau_2) \geq \epsilon$ for some threshold value ϵ. This leaves us with the problem of selecting a threshold for which there is no obvious choice without a significant calibration effort.

One popular approach is to create a series of M *surrogate* signals $s_m(t_{k_n})$ $n = 1 \cdots N$, $m = 1 \cdots M$. Each surrogate is a *phase-randomized* version of the original recorded response $x(t_n)$. That is to say, for each m, we form $s_m(t_{k_n}) = x(t_n)$ where the time indices k_n are chosen so that $R_{S_m S_m}(\tau) = R_{XX}(\tau)$. Since the surrogates are just temporally shuffled versions of the original signal, the marginal distributions are identical, i.e., $p_S(s) = p_X(x)$. However, because of the randomization (temporal shuffling) all other correlations, e.g., $C_{XXX}(\tau_1, \tau_2)$, are destroyed. If done correctly, the resulting surrogate signals are consistent with different realizations of a linear system response [43]. Thus, we may compute the function $C_{XXX}(\tau_1, \tau_2)$ on both data and surrogates and look for differences at a prescribed confidence level.

Of course, there may be some other function (other than $C_{XXX}(\tau_1, \tau_2)$) that reflects higher-order joint moments would be more sensitive to their presence. In this regard we have found the TDTE to be quite useful. Because it is defined in terms of the entire joint probability distribution of the data, it captures *all* higher-order correlations (not just third order). As a result, the TDTE is a good indicator of nonlinearity in system response data.

6.3.2.2   Numerical Example

As an example, returning to the linear system (6.55), assume that there has been inserted a nonlinear restoring force between masses 2 and 3. This restoring force can be written as an additional term on the right-hand side of (6.55)

$$
\mathbf{f}^{(N)} = \begin{bmatrix} 0 \\ -\mu k_3 (x_3 - x_2)^3 \\ \mu k_3 (x_3 - x_2)^3 \\ 0 \\ 0 \end{bmatrix}.
$$

This particular form of nonlinearity is sometimes used to model buckling in structural systems. The equilibrium point $x_3 - x_2 = 0$ is replaced by the two stable points $x_3 - x_2 = \pm\sqrt{1/\mu}$. As μ is increased, the asymmetry in restoring force associated with the nonlinearity also increases. For a large enough value $\mu = \mu^*$ this system will begin to oscillate between the two equilibria.

As a test statistic we consider the time-delayed transfer entropy and examine the coupling between x_3 to x_2, that is, compute $TE_{3\to2}$. As with mutual information, results can be compared to those obtained from surrogate data as nonlinearity is introduced into the system. Figure 6.14 illustrates these results for several different values of the nonlinearity parameter μ. The differences between surrogates and data are noticeable over a wider range of delay T. We may quantify this difference by forming the nonlinearity index

$$
Z_T = \sum_T \begin{cases} 0 &:& TE_{j\to i}(T) \le CU(T) \\ (TE_{j\to i}(T) - CU(T))/CU(T) &:& TE_{j\to i}(T) > CU(T) \end{cases}.
$$

which simply sums, over all delays, the degree to which the TE exceeds the confidence limit defined by the surrogates. Here we base the upper confidence limit $CU(T)$ on the mean and standard deviation of surrogate values for delay T, specifically $CU(T) = \mu_s(T) + 1.96\sigma_s(T)$. The values of this index are larger than are those for the mutual information, and many other nonlinearity indicators we have examined. We conclude that for this simple system, transfer entropy is a sensitive indicator of nonlinear coupling.

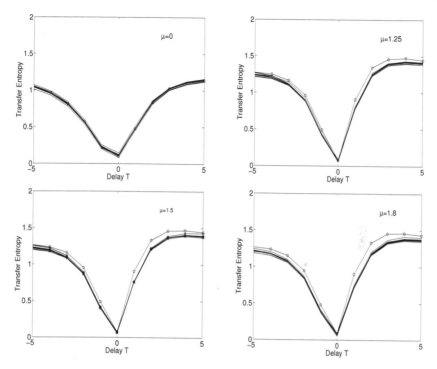

FIGURE 6.14

Plot of $TE_{3\to2}$ showing increasing discrepancy between surrogates (solid line) and data (open circles) for $\mu = 0$ (linear system), $\mu = 1.25$, $\mu = 1.5$, and $\mu = 1.75$.

6.3.2.3 Experimental Application: Diagnosing Rotor-Stator Rub

In rotary systems the owner/operator is often interested in diagnosing the presence of a rub between the shaft and the housing (stator). The undamaged rotor system is often modeled as linear with constant coefficient mass, stiffness, and damping matrices much like Eqn. (6.55) [26]. In this case the presence of a rotor-stator rub is expected to produce a nonlinearity (stick-slip/impacting) in an otherwise linear system. Thus, diagnosis of the rub may be performed in the absence of a baseline (no-rub) dataset using the transfer entropy in conjunction with the surrogate data method described above.

The system used in this experiment, depicted in Figure 6.16, consists of an Active Magnetic Bearing (AMB) (Revolve Magnetic Bearing, Inc.), inboard and outboard bearings (located at either end of the shaft), and a balance disk. The system also includes a PID controller for minimizing shaft position errors. Additionally, two proximity probes (Bently-Nevada®) for monitoring the shaft vibration were positioned at approximately the 1/4 and 3/4 points along the length of the shaft. Gaussian excitation was provided via the AMB located

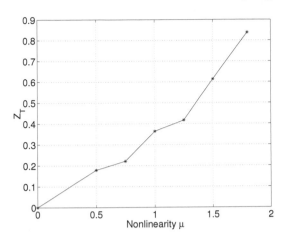

FIGURE 6.15
Index Z_T as a function of % nonlinearity (damage) in the system.

near the mid-span of the shaft. AMBs can be used to excite the shaft with arbitrary waveforms and thus provide a convenient mechanism for performing online damage detection. Data were collected from the pristine structure (no rub) and for two different rub scenarios. A total of 40 surrogates were generated from the original data, each consisting of $N = 32768$ points. The transfer entropy estimator was applied using data recorded from the inboard and outboard horizontal probes using a fixed-mass kernel with $M = 40$ points. Confidence intervals, based on the surrogates, were formed by discarding the low and high values of the surrogate data to give an interval of 95% confidence. These results are shown in Figure 6.17. For the healthy system the TDTE appears consistent with that of a linear system. As the damage is introduced, the TDTE values begin to fall outside the confidence intervals formed from the surrogates. As the severity of the rub is increased, the separation between surrogates and original data increases as can be seen from the lower plots in Figure 6.17. Damage in this case clearly manifests itself as the presence of a nonlinearity. The TDTE values across nearly all delays exceed the confidence limits for both rub scenarios. The strength of this approach is that each of the plots in Figure 6.17 can be viewed independently to assess the level of damage. Each is providing an absolute measure of nonlinearity (damage); hence, training data from the healthy system is not needed.

FIGURE 6.16
Overview of rotor experiment (left) and close up of "rub" screw (right).

6.4 Summary

The goal of this book is to provide a comprehensive overview of the differential entropy and to at least partially demonstrate its applicability to problems in engineering and science. This chapter has focused on the latter issue. Both the time-delayed mutual information and time-delayed transfer entropy functions were shown to be comprised of continuous differential entropies. These measures both quantify the relationship between two or more random variables, but do so in very different ways. The mutual information simply captures deviations from the hypothesis that two random variables are statistically independent. We have shown that by adding a time delay, this measure can be used to provide maximum likelihood estimates of a time delay between two radar pulses regardless of the corrupting noise model. In fact, for highly non-Gaussian noise models the time-delayed mutual information can significantly outperform the linear cross-correlation in estimating the delay time. The transfer entropy quantifies a degree of dynamical dependence between two random processes. We have shown how this measure can be used to better understand coupling in linear systems and also to detect the presence of nonlinearity in a dynamical system. While other uses of both information-theoretics have been developed, the examples provided here give a flavor for the types of problems that can be aided by an understanding of the differential entropy.

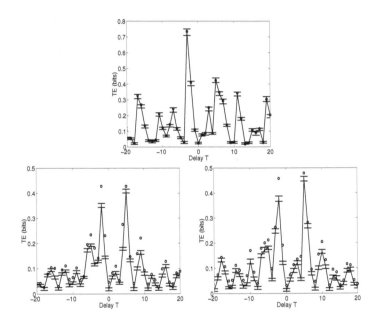

FIGURE 6.17

Time-delay transfer entropy computed from original data (open circles) compared to values obtained from surrogates for undamaged (top) and damaged (bottom plots) rotor.

7

Appendices

7.1 Derivation of Maximum Entropy Distributions under Different Constraints

Proof that the Uniform Distribution has Maximum Entropy for Discrete Variables Subject Only to the Normalization Constraint

Given a discrete random variable X having N possible outcomes and governed by a probability mass function $f_X(x_i)$, we wish to prove that the Uniform distribution

$$f_X(x_i) = 1/N \ \forall \ i \in \{1, 2, \cdots, N\} \tag{7.1}$$

is the distribution that maximizes the discrete Shannon entropy

$$H_X = -\sum_{i=1}^{N} f_X(x_i) \log_2 \left(f_X(x_i) \right) \tag{7.2}$$

subject to the normalization constraint

$$\sum_{i=1}^{N} f_X(x_i) = 1. \tag{7.3}$$

We use the method of Lagrange multipliers. First, form the functional

$$G(f_i, \lambda) = H_X - \lambda \left(\sum_{i=1}^{N} f_i - 1 \right), \tag{7.4}$$

where we have used the simplified notation $f_i \equiv f_X(x_i)$, and then set all N derivatives equal to zero

$$\frac{\partial G(f_i, \lambda)}{\partial f_i} = \log_2 f_i + f_i(1/f_i) - \lambda = 0. \tag{7.5}$$

Hence, $\log_2 f_i = \lambda - 1$ for all i, or

$$f_i = 2^{\lambda - 1}. \tag{7.6}$$

Solve for λ by substituting back into the normalization constraint

$$\sum_{i=1}^{N} f_i = \frac{1}{2} \sum_{i=1}^{N} 2^\lambda = 1 \qquad (7.7)$$

from which we find $\lambda = \log_2 2 - \log_2 N$. Hence,

$$f_i = 2^{\lambda-1} = \frac{1}{N} \qquad (7.8)$$

which is just the Uniform distribution.

Proof that the Normal Distribution has Maximum Entropy for Continuous Variables on Infinite Support with Specified Variance

Given a continuous random variable X governed by a probability density $p_X(x)$, we wish to prove that the normal distribution

$$p_X(x) = \frac{1}{\sqrt{2\pi\sigma^2}} \exp\left(-\frac{x^2}{2\sigma^2}\right) \quad -\infty \le x \le \infty \qquad (7.9)$$

is the distribution that maximizes the differential Shannon entropy

$$h_X = -\int_{-\infty}^{\infty} p_X(x) \log_2\left(p_X(x)\right) dx. \qquad (7.10)$$

subject to the two constraints

$$\int_{-\infty}^{\infty} p_X(x) dx = 1$$

$$\int_{-\infty}^{\infty} x^2 p_X(x) dx = \sigma^2. \qquad (7.11)$$

Again, we use the method of Lagrange multipliers. First, form the functional

$$G(p; \lambda, \eta) = h_X - \lambda \left(\int p \, dx - 1\right) - \eta \left(\int x^2 p \, dx - \sigma^2\right), \qquad (7.12)$$

where we have used the simplified notation $p \equiv p_X(x)$, and set its derivative equal to zero,

$$\frac{\partial G(p; \lambda, \eta)}{\partial p} = \int \left(-\ln p - 1 - \lambda - \eta x^2\right) dx = 0. \qquad (7.13)$$

Note that we have taken the natural log here to ease the calculation. In order for the integral to be zero for arbitrary p the integrand itself must be identically zero

$$- \ln p - 1 - \lambda - \eta x^2 = 0 \qquad (7.14)$$

leading to $\ln p = -(\lambda + 1) - \eta x^2$ or

$$p = e^{-(\lambda+1)} e^{-\eta x^2}. \qquad (7.15)$$

To obtain λ and η, insert p back into the constraints

$$\int_{-\infty}^{\infty} e^{-(\lambda+1)} e^{-\eta x^2} dx = 1$$

$$\int_{-\infty}^{\infty} x^2 e^{-(\lambda+1)} e^{-\eta x^2} dx = \sigma^2. \qquad (7.16)$$

Evaluation of the first constraint yields $e^{-(\lambda+1)} = \sqrt{\eta/\pi}$. Substituting into the second constraint and evaluating the integral leads to $\eta = 1/2\sigma^2$. So, finally,

$$p = e^{-(\lambda+1)} e^{-\mu x^2} = \frac{1}{\sqrt{2\pi\sigma^2}} \exp\left(-x^2/2\sigma^2\right) \qquad (7.17)$$

7.2 Moments and Characteristic Function for the Sine Wave Distribution

The rth Moment about the origin is given by:

$$\mu'_r = \frac{1}{\pi} \int_{-A}^{A} x^r \frac{1}{\sqrt{A^2 - x^2}} dx$$

$$= \begin{cases} 0 & : \quad r \quad \text{odd} \\ \frac{2}{\pi} \int_0^A \frac{x^r}{\sqrt{A^2 - x^2}} dx & : \quad r \quad \text{even} \end{cases}$$

In the case where r is even, let $y = \frac{x}{A}$ so that $x = Ay$ and $\sqrt{A^2 - x^2} = A\sqrt{1 - y^2}$. Then

$$\mu'_r = \frac{2}{\pi} \int_0^1 \frac{A^r y^r}{A\sqrt{1-y^2}} A dy$$

$$= \frac{2A^r}{\pi} \int_0^1 \frac{y^r}{\sqrt{1-y^2}} dy$$

$$= \frac{2A^r}{\pi} \left(\frac{1 \cdot 3 \cdot 5 \cdots (r-1)}{2 \cdot 4 \cdots r} \frac{\pi}{2} \right)$$

See Gradshteyn & Ryzhik, 3.248(3) for this last step. Note that in their notation $(2n-1)!!$ is used to represent the product of odd integers while $(2n)!!$ does the same for even integers. So

$$\mu'_r = \begin{cases} 0 & : \text{ if } \quad r \text{ odd} \\ A^r \left(\frac{1 \cdot 3 \cdot 5 \cdots (r-1)}{2 \cdot 4 \cdots r} \right) & : \quad r \text{ even} \end{cases}.$$

In particular

$$\mu'_2 = A^2 \left(\frac{1}{2} \right) = \frac{A^2}{2}$$

and, since the mean $\mu'_1 = 0$, the variance is also $\frac{A^2}{2}$.

The characteristic function for this distribution is

$$E[e^{itx}] = \frac{1}{\pi} \int_{-A}^{A} \frac{e^{itx}}{\sqrt{A^2 - x^2}} dx$$

$$= \frac{1}{\pi} \left[\int_{-A}^{A} \frac{\cos(tx)}{\sqrt{A^2 - x^2}} dx + i \int_{-A}^{A} \frac{\sin(tx)}{\sqrt{A^2 - x^2}} dx \right]$$

$$= \frac{2}{\pi} \int_0^A \frac{\cos(tx)}{\sqrt{A^2 - x^2}} dx$$

since the first integrand is even and the second is odd. As before, let $y = x/A$ so that $x = Ay$ and $\sqrt{A^2 - x^2} = A\sqrt{1 - y^2}$.

$$= \frac{2}{\pi} \int_0^1 \frac{\cos(Aty)}{A\sqrt{1-y^2}} A dy$$

$$= \frac{2}{\pi} \int_0^1 \frac{\cos(Aty)}{\sqrt{1-y^2}} dy$$

where from Gradshteyn & Ryzhik 3.753(2) we have

$$= \frac{2}{\pi}\frac{\pi}{2} J_o(At)$$
$$= J_o(At)$$

where $J_o(\cdot)$ is the 0th order Bessel function of the first kind.

7.3 Moments, Mode, and Characteristic Function for the Mixed-Gaussian Distribution

For the mixed-Gaussian distribution the rth moment about the origin is equal to $\frac{1}{2}\times$ the rth moment about the origin for a Gaussian distributed random variable centered at $+\mu$ plus $\frac{1}{2}\times$ the rth moment about the origin for a Gaussian distributed random variable centered at $-\mu$.

$$\mu'_r = \frac{1}{2}\left[\mu^r + \sum_{\substack{k=2\\ k\ even}}^{r} \binom{r}{k}\mu^{r-k}\frac{\sigma^k k!}{2^{k/2}(k/2)!}\right]$$

$$+ \frac{1}{2}(-1)^r\left[\mu^r + \sum_{\substack{k=2\\ k\ even}}^{r} \binom{r}{k}(-1)^r(-1)^{-k}\mu^{r-k}\frac{\sigma^k k!}{2^{k/2}(k/2)!}\right]$$

$$= \frac{1}{2}\left[\mu^r + \sum_{\substack{k=2\\ k\ even}}^{r} \binom{r}{k}\mu^{r-k}\frac{\sigma^k k!}{2^{k/2}(k/2)!}\right]$$

$$+ \frac{1}{2}(-1)^r\left[\mu^r + \sum_{\substack{k=2\\ k\ even}}^{r} \binom{r}{k}\mu^{r-k}\frac{\sigma^k k!}{2^{k/2}(k/2)!}\right]$$

$$= \begin{cases} 0 & : \quad r \quad \text{odd} \\ \mu^r + \sum_{\substack{k=2\\ k\ even}}^{r} \binom{r}{k}\mu^{r-k}\frac{\sigma^k k!}{2^{k/2}(k/2)!} & : \quad r \quad \text{even} \end{cases}$$

which is the same as the rth moment about the origin for a Gaussian random variable with mean μ. In particular

Mean= 0

Variance=second moment about the origin=second moment about the origin of a Gaussian with mean $\mu = \sigma^2 + \mu^2$.

The mode for the mixed-Gaussian distribution can be obtained by noting that the distribution becomes bi-modal when the value of the Gaussian centered at μ at $x = 0$ becomes less than $\frac{1}{2}$ the peak value of $\frac{1}{\sigma\sqrt{2\pi}}$ at $x = \mu$.

This occurs when

$$\frac{1}{2}\frac{1}{\sigma\sqrt{2\pi}} = \frac{1}{\sigma\sqrt{2\pi}}e^{-(x-\mu)^2/2\sigma^2}$$

$$\frac{1}{2} = e^{-(x-\mu)^2/2\sigma^2}$$

$$-\ln(2) = -\frac{(x-\mu)^2}{2\sigma^2}$$

$$(x-\mu)^2 = 2\sigma^2\ln(2) = \sigma^2\ln(4)$$

$$x = \mu \pm \sigma\sqrt{\ln(4)}$$

So the distribution becomes bi-modal exactly when $\mu - \sigma\sqrt{\ln(4)} > 0$ or $\mu > \sigma\sqrt{\ln(4)} = 1.1774\sigma$.

The Characteristic Function is given by:

$$E[e^{itx}] = \frac{1}{2}E[e^{itx}] \quad \text{for Gaussian with mean } \mu$$

$$+ \frac{1}{2}E[e^{itx}] \quad \text{for Gaussian with mean } -\mu$$

$$= \frac{1}{2}\left[e^{i\mu t - \frac{1}{2}\sigma^2 t^2} + e^{-i\mu t - \frac{1}{2}\sigma^2 t^2}\right]$$

$$= \frac{1}{2}e^{-\sigma^2 t^2/2}\left(e^{i\mu t} + e^{-i\mu t}\right)$$

$$= e^{-\sigma^2 t^2/2}\cos(\mu t).$$

7.4 Derivation of Function $L(\alpha)$ Used in Derivation for Entropy of Mixed-Gaussian Distribution

An analytic expression for the integral in Eqn. (4.4) in the derivation of the differential entropy for the Mixed-Gaussian distribution could not be found. However, there are analytic bounds for the integral term which are derived by noting that

$$y - \ln 2 \le \ln(\cosh(y)) \le y \quad \forall \ y \ge 0.$$

Thus, for the upper bound to the integral term we have

$$I = \frac{2}{\sqrt{2\pi}\alpha} e^{-\alpha^2/2} \int_0^\infty e^{-y^2/2\alpha^2} \cosh(y) \ln(\cosh(y)) dy$$

$$\leq \frac{2}{\sqrt{2\pi}\alpha} e^{-\alpha^2/2} \int_0^\infty e^{-y^2/2\alpha^2} \cosh(y) y \, dy$$

$$= \frac{2}{\sqrt{2\pi}\alpha} e^{-\alpha^2/2} \left[\frac{\alpha^2}{2} \sqrt{2\alpha^2\pi} e^{\alpha^2/2} \operatorname{erf}(\alpha/\sqrt{2}) + \alpha^2 \right]$$

$$= \alpha^2 \operatorname{erf}(\alpha/\sqrt{2}) + \sqrt{2/\pi}\alpha e^{-\alpha^2/2} \tag{7.18}$$

by means of formula 3.562 (4) in [14], where erf denotes the error function, defined as

$$\operatorname{erf}(z) = \frac{2}{\sqrt{\pi}} \int_0^z e^{-u^2} du. \tag{7.19}$$

Likewise, for the lower bound we have

$$I \geq \alpha^2 \operatorname{erf}(\alpha/\sqrt{2}) + \sqrt{2/\pi}\alpha e^{-\alpha^2/2} - \frac{2}{\sqrt{2\pi}\alpha} e^{-\alpha^2/2} \ln 2 \int_0^\infty e^{-y^2/2\alpha^2} \cosh(y) dy$$

$$= \alpha^2 \operatorname{erf}(\alpha/\sqrt{2}) + \sqrt{2/\pi}\alpha e^{-\alpha^2/2} - \frac{2}{\sqrt{2\pi}\alpha} e^{-\alpha^2/2} \ln 2 \left[\frac{1}{2} \sqrt{2\alpha^2\pi} e^{\alpha^2/2} \right]$$

$$= \alpha^2 \operatorname{erf}(\alpha/\sqrt{2}) + \sqrt{2/\pi}\alpha e^{-\alpha^2/2} - \ln 2 \tag{7.20}$$

by means of formula 3.546 (2) in [14].

Since the integrand in I is always greater than or equal to 0, we know that $I \geq 0$, so we can write

$$h_e(X) = \frac{1}{2} \ln(2\pi e\sigma^2) + \alpha^2 - I \tag{7.21}$$

where

$$\max(0; \alpha^2 \operatorname{erf}(\alpha/\sqrt{2}) + \sqrt{2/\pi}\alpha e^{-\alpha^2/2} - \ln 2) \leq I \leq \alpha^2 \operatorname{erf}(\alpha/\sqrt{2}) + \sqrt{2/\pi}\alpha e^{-\alpha^2/2}$$

for all $\alpha = \mu/\sigma \geq 0$. The graph of I as a function of α is shown in Figure 7.1, along with the analytic upper and lower bounds. Clearly I converges rapidly to the lower bound as α increases. A tabulation of numerically computed values of I is presented in Table 7.1, together with corresponding values of $\alpha^2 - I$. As is clear in the table, $(\alpha^2 - I)$ monotonically increases from 0 to $\ln 2 = 0.6931$. Hence the differential entropy, in nats, of a mixed-Gaussian distribution can be expressed as

$$h_e(X) = \frac{1}{2} \ln(2\pi e\sigma^2) + (\alpha^2 - I) \tag{7.22}$$

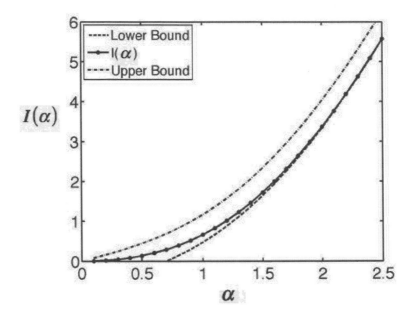

FIGURE 7.1
Lower and upper bounds for $I(\alpha)$ vs. α.

where $(\alpha^2 - I)$ is a function of $\alpha = \mu/\sigma$ which is equal to zero at $\alpha = 0$ (in which case the distribution is Gaussian) and monotonically increases to $\ln 2$ as α increases to $\alpha > 3.5$ (in which case the distribution is effectively split into two separate Gaussians). In particular, if $\sigma = 1$, $h_e(X)$ is a monotonically increasing function of μ which has the value 1.419 for $\mu = 0$ and converges to the value 2.112 as μ is increased and the two parts of the mixed-Gaussian distribution are split apart.

To express the differential entropy in bits, Eqn. (7.22) needs to be divided by $\ln 2$, which gives

$$h(x) = \frac{1}{2}\log_2(2\pi e\sigma^2) + \left(\frac{\alpha^2 - I}{\ln 2}\right) \tag{7.23}$$

where the second term, which is denoted by $L(\alpha)$, is a monotonically increasing function of $\alpha = \mu/\sigma$ which goes from 0 at $\alpha = 0$ to 1 for $\alpha > 3.5$ (see Table 7.2). In particular, for $\sigma = 1$, the differential entropy in bits goes from 2.05 to 3.05 depending on the value of μ; that is, depending on how far apart the two halves of the mixed-Gaussian distribution are.

TABLE 7.1
Tabulated values for $I(\alpha)$ and $\alpha^2 - I$.

α	$I(\alpha)$	$\alpha^2 - I$	α	$I(\alpha)$	$\alpha^2 - I$
0.0	0.000	0.000	(continued)		
0.1	0.005	0.005	2.1	3.765	0.645
0.2	0.020	0.015	2.2	4.185	0.656
0.3	0.047	0.043	2.3	4.626	0.664
0.4	0.086	0.074	2.4	5.089	0.671
0.5	0.139	0.111	2.5	5.574	0.676
0.6	0.207	0.153	2.6	6.080	0.680
0.7	0.292	0.198	2.7	6.607	0.683
0.8	0.396	0.244	2.8	7.154	0.686
0.9	0.519	0.291	2.9	7.722	0.688
1.0	0.663	0.337	3.0	8.311	0.689
1.1	0.829	0.381	3.1	8.920	0.690
1.2	1.018	0.422	3.2	9.549	0.691
1.3	1.230	0.460	3.3	10.198	0.692
1.4	1.465	0.495	3.4	10.868	0.692
1.5	1.723	0.527	3.5	11.558	0.692
1.6	2.005	0.555	3.6	12.267	0.693
1.7	2.311	0.579	3.7	12.997	0.693
1.8	2.640	0.600	3.8	13.747	0.693
1.9	2.992	0.618	3.9	14.517	0.693
2.0	3.367	0.633	4.0	15.307	0.693

TABLE 7.2

Tabulated values for $L(\alpha)$.

α	$L(\alpha)$	α	$L(\alpha)$
0.0	0.000	(continued)	
0.1	0.007	2.1	0.931
0.2	0.028	2.2	0.946
0.3	0.062	2.3	0.958
0.4	0.107	2.4	0.968
0.5	0.161	2.5	0.975
0.6	0.221	2.6	0.981
0.7	0.286	2.7	0.986
0.8	0.353	2.8	0.990
0.9	0.420	2.9	0.992
1.0	0.486	3.0	0.994
1.1	0.549	3.1	0.996
1.2	0.609	3.2	0.997
1.3	0.664	3.3	0.998
1.4	0.714	3.4	0.999
1.5	0.760	3.5	0.999
1.6	0.800	3.6	0.999
1.7	0.835	3.7	1.000
1.8	0.865	3.8	1.000
1.9	0.891	3.9	1.000
2.0	0.913	4.0	1.000

7.5   References to Formulae Used in This Text

TABLE 7.3
Integrals used from Gradshteyn and Ryzhik [14].

page	number	formula
285	3.194(3)	$\int_0^\infty \frac{x^{\mu-1}dx}{(1+\beta x)^\nu} = \beta^{-\mu}B(\mu, \nu-\mu)$ $[|arg\ \beta| < \pi,\ Re(\nu) > Re(\mu) > 0]$
294	3.248(3)	$\int_0^1 \frac{x^{2n}dx}{\sqrt{1-x^2}} = \frac{(2n-1)!!}{(2n)!!}\frac{\pi}{2}$
342	3.478(1)	$\int_0^\infty x^{\nu-1}\exp(-\mu x^P)dx = \frac{1}{P}\mu^{-\nu/P}\Gamma\left(\frac{\nu}{P}\right)$ $[Re(\mu) > 0,\ Re(\nu) > 0,\ P > 0]$
419	3.753(2)	$\int_0^1 \frac{\cos(ax)dx}{\sqrt{1-x^2}} = \frac{\pi}{2}J_o(a)$
532	4.231(6)	$\int_0^1 \frac{\ln(x)}{(1+x)^2}dx = -\ln(2)$
535	4.241(7)	$\int_0^1 \frac{\ln(x)}{\sqrt{1-x^2}}dx = -\frac{\pi}{2}\ln(2)$
538	4.253(1)	$\int_0^1 x^{\mu-1}(1-x^r)^{\nu-1}\ln(x)dx = \frac{1}{r^2}B\left(\frac{\mu}{r}, \nu\right) \times$ $\left[\Psi\left(\frac{\mu}{r}\right) - \Psi\left(\frac{\mu}{r}+\nu\right)\right][Re(\mu) > 0,\ Re(\nu) > 0,\ r > 0]$
538	4.253(3)	$\int_u^\infty \frac{(x-u)^{\mu-1}\ln(x)dx}{x^\lambda} = u^{\mu-\lambda}B(\lambda-\mu, \mu)\times$ $[\ln(u) + \Psi(\lambda) - \Psi(\lambda-\mu)][0 < Re(\mu) < Re(\lambda)]$
558	4.293(14)	$\int_0^\infty \frac{x^{\mu-1}\ln(\gamma+x)dx}{(\gamma+x)^\nu} = \gamma^{\mu-\nu}B(\mu, \nu-\mu)\times$ $[\Psi(\nu) - \Psi(\nu-\mu) + \ln(\gamma)][0 < Re(\mu) < Re(\nu)]$

Prudnikov Vol. 1 [44]

$$\int_0^\infty \frac{x^{\alpha-1}\ln(x)}{(x^\mu+z^\mu)^\rho}dx = \frac{z^{\alpha-\rho\mu}}{\mu^2}B\left(\frac{\alpha}{\mu}, \rho-\frac{\alpha}{\mu}\right)\left[\mu\ln(z) + \Psi\left(\frac{\alpha}{\mu}\right) - \Psi\left(\rho-\frac{\alpha}{\mu}\right)\right]$$

$$[\mu > 0,\ \alpha < \mu\rho]$$

TABLE 7.4
Integrals used from Korn and Korn [22].

page	number	formula
331	39	$\int_0^\infty x^b e^{-ax^2} dx = \frac{\Gamma\left(\frac{b+1}{2}\right)}{2a^{(b+1)/2}}$ *(for $a > 0$, $b > -1$)*
		$= \begin{cases} \frac{1 \cdot 3 \cdots (b-1)\sqrt{\pi}}{2^{b/2+1} a^{(b+1)/2}} & \textit{(for } a > 0, \ b = 0, 2, 4, \cdots) \\ \frac{\left(\frac{b-1}{2}\right)!}{2a^{(b+1)/2}} & \textit{(for } a > 0, \ b = 1, 3, 5, \cdots) \end{cases}$
331	42	$\int_0^\infty x^2 e^{-x^2} dx = \frac{\sqrt{\pi}}{4}$
332	61	$\int_0^\infty e^{-x} \ln(x) dx = -\gamma$ (Euler-Mascheroni constant)

7.6   Useful Theorems

Theorem 1 *(Isserlis' Theorem) If $\eta_1, \eta_2, \cdots, \eta_{2N+1}$ $(N = 1, 2, \cdots)$ are normalized, jointly Gaussian random variables (i.e., for every i, $E[\eta_i] = 0$ and $E[\eta_i^2] = 1$), then*

$$E[\eta_1 \eta_2 \cdots \eta_{2N}] = \sum \prod E[\eta_i \eta_j]$$

and

$$E[\eta_1 \eta_2 \cdots \eta_{2N+1}] = 0$$

where the notation $\sum \prod$ means summing over all distinct ways of partitioning $\eta_1, \eta_2, \cdots, \eta_{2N}$ into pairs. That is, the $2N$ variables $\eta_1, \eta_2, \cdots, \eta_{2N}$ are partitioned into cells containing pairs of the variables. The expected values $E[\eta_i \eta_j]$ are formed for the products of the pairs in each cell, and these expected values are multiplied together for all N cells. These products are then summed for all partitions into cells which are distinct; that is, the ordering of the cells and of the pairs within the cells is not distinguished. Therefore, there are

$$\frac{(2N)!}{2^N N!} \tag{7.24}$$

terms in the $\sum \prod$ expression for $2N$ variables. For example, consider the case of four Gaussian random variables $\eta_1, \eta_2, \eta_3, \eta_4$. Application of the above theorem gives the relation $E[\eta_1 \eta_2 \eta_3 \eta_4] = E[\eta_1 \eta_2] E[\eta_3 \eta_4] + E[\eta_1 \eta_3] E[\eta_2 \eta_4] + E[\eta_1 \eta_4] E[\eta_2 \eta_3]$. Here there are only three terms, but the number of terms increases rapidly as the number of variables increases; for example, there are 135,135 terms for 14 variables.

Theorem 2 *(Isserlis' Theorem for Mixed-Gaussian random variables [34]) If $x_1, x_2, \cdots, x_{2N}$ are jointly mixed-Gaussian random variables with parameters μ_i, σ_i, then*

$$
\begin{aligned}
E[x_1 x_2 x_3 x_4 \cdots x_{2N}] = &\sum \prod^{2N} E[\zeta_i \zeta_j] \\
&+ \sum_{k=1}^{N-1} \left(\sum_{i_1 < i_2 < \cdots < i_{2k}}^{2NC_{2k}} \mu_{i_1} \mu_{i_2} \cdots \mu_{i_{2k}} \left(\sum \prod^{2N-2k} E[\zeta_i \zeta_j] \right) \right) \\
&+ \mu_1 \mu_2 \cdots \mu_{2N}
\end{aligned}
$$
(7.25)

$$
E[x_1 x_2 x_3 x_4 \cdots x_{2N-1}] = 0
$$
(7.26)

Here the averages on the left-hand side of the equation are over the $2N$-dimensional (or (2N-1)-dimensional) jointly mixed-Gaussian distribution, while the averages on the right-hand side are all over the two-dimensional jointly Gaussian distributions where $\zeta_1, \zeta_2, \cdots, \zeta_{2N}$ are jointly Gaussian random variables with zero means and standard deviations σ_i.

In the above notation, $\sum \prod^{2} E[\zeta_i \zeta_j]$ stands for $E[\zeta_i \zeta_j]$ with $i < j$. $\sum \prod^{2N}$ is the sum over all distinct ways of partitioning $\zeta_1, \zeta_2, \cdots, \zeta_{2N}$ into pairs. $\sum \prod^{2N-2k}$ is the sum over all distinct ways of partitioning $\zeta_{j_1}, \zeta_{j_2}, \cdots, \zeta_{j_{2N-2k}}$ into pairs where $j_1 < j_2 < \cdots < j_{2N-2k}$ and $\{i_1, i_2, \cdots, i_{2k}\} \cup \{j_1, j_2, \cdots, j_{2N-2k}\} = \{1, 2, \cdots, 2N\}$.

Bibliography

[1] M. Abramowitz and I. A. Stegun. *Handbook of Mathematical Functions.* Dover Publications, New York, 1965.

[2] M. Bauer, J. W. Cox, M. H. Caveness, J. J. Downs, and N. F. Thornhill. Finding the direction of disturbance propagation in a chemical process using transfer entropy. *IEEE Transactions on Control Systems Technology*, 15(1):12–21, 2007.

[3] H. Benaroya. *Mechanical Vibration: Analysis, Uncertainties, and Control.* Prentice Hall, New Jersey, 1998.

[4] J. S. Bendat and A. G. Piersol. *Random Data Analysis and Measurement Procedures, Third Edition.* Wiley & Sons, New York, NY, 2000.

[5] J. Benetsy, Y. Huang, and J. Chen. Time delay estimation via minimum entropy. *IEEE Signal Processing Letters*, 14(3):157–160, 2007.

[6] G. D. Birkhoff. Proof of the ergodic theorem. *Proceedings of the National Academy of Sciences*, 17:656–660, 1931.

[7] H. B. Callen. *Thermodynamics.* John Wiley & Sons, New York, 1960.

[8] T. M. Cover and J. A. Thomas. *Elements of Information Theory.* John Wiley & Sons, New Jersey, 2006.

[9] S. H. Crandall and W. D. Mark. *Random Vibration in Mechanical Systems.* Academic Press, New York, 1963.

[10] N. Ebrahimi, E. Maasoumi, and E. S. Soofi. Ordering univariate distributions by entropy and variance. *Journal of Econometrics*, 90:317–336, 1999.

[11] D. Erdogmus, R. Agrawal, and J. C. Principe. A mutual information extension to the matched filter. *Signal Processing*, 85(5):927–935, 2005.

[12] R. J. Evans, J. Boersma, N. M. Blachman, and A. A. Jagers. The entropy of a poisson distribution: Problem 87-6. *SIAM Review*, 30(2):314–317, 1988.

[13] R. P. Feynman and A. Hey. *Feynman Lectures on Computation.* Westview Press, Boulder, CO, 1996.

[14] I. S. Gradshteyn and I. M. Ryzhik. *Table of Integrals, Series and Products, 4th ed.* 6th printing, Academic Press, New York, 1980.

[15] D. W. Hahs and S. D. Pethel. Distinguishing anticipation from causality: Anticipatory bias in the estimation of information flow. *Physical Review Letters*, 107:128–701, 2011.

[16] M. Henon. On the numerical computation of Poincaré maps. *Physica D*, 5:412–414, 1982.

[17] E. T. Jaynes. Information theory and statistical mechanics. *Physical Review*, 106:620–630, 1957.

[18] E. T. Jaynes. *Probability Theory: The Logic of Science.* Cambridge University Press, New York, 2003.

[19] A. Kaiser and T. Schreiber. Information transfer in continuous processes. *Physica D*, 166:43–62, 2002.

[20] S. M. Kay. *Fundamentals of Statistical Signal Processing: Volume I, Estimation Theory.* Prentice Hall, New Jersey, 1993.

[21] A. N. Kolmogorov. *Foundations of Probability.* Chelsea Publishing Co., New York, 1956.

[22] G. A. Korn and T. M. Korn. *Manual of Mathematics.* McGraw-Hill, New York, 1967.

[23] S. Kullback. *Information Theory and Statistics.* Dover Publications, Inc., New York, 1968.

[24] A. C. G. Verdugo Lazo and P. N. Rathie. On the entropy of continous probability distributions. *IEEE Transactions on Information Theory*, IT-24(1):120–122, 1978.

[25] W. A. Link and R. J. Barker. *Bayesian Inference with ecological examples.* Academic Press, San Diego, CA, 2010.

[26] G. Mani, D. D. Quinn, and M. Kasarda. Active health monitoring in a rotating cracked shaft using active magnetic bearings as force actuators. *Journal of Sound and Vibration*, 294(3):454–465, 2006.

[27] N.J.I. Mars and G.W. Van Arragon. Time delay estimation in nonlinear systems. *IEEE Transactions on Acoustics, Speech, and Signal Processing*, 29(3):619–621, 1981.

[28] R. Marschinski and H. Kantz. Analysing the information flow between financial time series: an improved estimator for transfer entropy. *European Physics Journal*, B, 30:275–281, 2002.

[29] R. N. McDonough and A. D. Whalen. *Detection of Signals in Noise, second edition.* Academic Press, San Diego, 1995.

[30] L. Meirovitch. *Introduction to Dynamics and Control.* Wiley & Sons, New York, 1985.

[31] J. V. Michalowicz. Mathematical analysis of the counterfire duel: Tanks vs. anti-tank munitions. *Mathematical Modelling,* 5:23–42, 1984.

[32] J. V. Michalowicz, J. M. Nichols, and F. Bucholtz. Calculation of differential entropy for a mixed Gaussian distribution. *Entropy,* 10:200–206, 2009.

[33] J. V. Michalowicz, J. M. Nichols, and F. Bucholtz. Calculation of entropy and mutual information for sinusoids. Technical Report NRL/MR/5650–09–, Naval Research Laboratory, 2009.

[34] J. V. Michalowicz, J. M. Nichols, F. Bucholtz, and C. C. Olson. A general Isserlis theorem for mixed-Gaussian random variables. *Statistics and Probability Letters,* 81(8):1233–1240, 2011.

[35] L. J. Moniz, E. G. Cooch, S. P. Ellner, J. D. Nichols, and J. M. Nichols. Application of information theory methods to food web reconstruction. *Ecological Modelling,* 208:145–158, 2007.

[36] L. J. Moniz, J. D. Nichols, and J. M. Nichols. Mapping the information landscape: Discerning peaks and valleys for ecological monitoring. *Journal of Biological Physics,* 33:171–181, 2007.

[37] A. M. Mood, F. A. Graybill, and D. C. Boes. *Introduction to the Theory of Statistics, 3rd edition.* McGraw-Hill, New York, 1974.

[38] J. M. Nichols. Inferences about information flow and dispersal for spatially extended population systems using time-series data. *Proceedings of the Royal Society of London - B series,* 272:871–876, 2005.

[39] J. M. Nichols. Examining structural dynamics using information flow. *Probabilistic Engineering Mechanics,* 21:420–433, 2006.

[40] J. M. Nichols, C. C. Olson, J. V. Michalowicz, and F. Bucholtz. The bispectrum and bicoherence for quadratically nonlinear systems subject to non-Gaussian inputs. *IEEE Transactions on Signal Processing,* 57(10):3879–3890, 2009.

[41] J. M. Nichols, M. Seaver, S. T. Trickey, L. W. Salvino, and D. L. Pecora. Detecting impact damage in experimental composite structures: an information-theoretic approach. *Smart Materials and Structures,* 15:424–434, 2006.

[42] J. M. Nichols, S. T. Trickey, and M. Seaver. Detecting damage-induced nonlinearities in structures using information theory. *Journal of Sound and Vibration*, 297:1–16, 2006.

[43] J. M. Nichols, S. T. Trickey, M. Seaver, S. R. Motley, and E. D. Eisner. Using ambient vibrations to detect loosening of a composite-to-metal bolted joint in the presence of strong temperature fluctuations. *Journal of Vibration and Acoustics*, 129:710–717, 2007.

[44] A. P. Prudnikov, Yu. A. Brychkov, and O. I. Marichev. *Integrals and Series, Volume I*. Gordon & Breach Science Publishers, New York, 1986.

[45] G. K. Rohde, J. M. Nichols, and F. Bucholtz. Pulse delay time estimation in non-Gaussian noise using mutual information maximization. *Signal Processing*, to appear, 2013.

[46] J. S. Rosenthal. *A first look at rigorous probability theory*. World Scientific, Singapore, 2000.

[47] T. Schreiber. Measuring information transfer. *Physical Review Letters*, 85:461, 2000.

[48] C. E. Shannon. A mathematical theory of communication. *The Bell System Technical Journal*, 27:379–423, 623–656, 1948.

[49] B. W. Silverman. *Density Estimation for Statistics and Data Analysis*. Chapman & Hall, London, 1986.

[50] I. Trendafilova, W. Heylen, and H. Van Brussel. Measurement point selection in damage detection usnig the mutual information concept. *Smart Materials and Structures*, 10(3):528–533, 2001.

[51] J. A. Vastano and H. L. Swinney. Information transport in spatiotemporal systems. *Physical Review Letters*, 60(18):1773–1776, 1988.

[52] M. W. Zemansky. *Heat and Thermodynamics, 5th edition*. McGraw-Hill, New York, 1968.

Index

Printed and bound by CPI Group (UK) Ltd, Croydon, CR0 4YY

21/10/2024

01777089-0007